Advanced Concrete Technology

Advanced Concrete Technology

Constituent Materials   ISBN 0 7506 5103 2

Concrete Properties   ISBN 0 7506 5104 0

Processes   ISBN 0 7506 5105 9

Testing and Quality   ISBN 0 7506 5106 7

Advanced Concrete Technology
Testing and Quality

Edited by
John Newman
Department of Civil Engineering
Imperial College
London

Ban Seng Choo
School of the Built Environment
Napier University
Edinburgh

ELSEVIER
BUTTERWORTH
HEINEMANN

AMSTERDAM BOSTON HEIDELBERG LONDON NEW YORK OXFORD
PARIS SAN DIEGO SAN FRANCISCO SINGAPORE SYDNEY TOKYO

Butterworth-Heinemann
An imprint of Elsevier
Linacre House, Jordan Hill, Oxford OX2 8DP
200 Wheeler Road, Burlington MA 01803

First published 2003

Copyright © 2003, Elsevier Ltd. All rights reserved

No part of this publication may be reproduced in any material
form (including photocopying or storing in any medium by
electronic means and whether or not transiently or incidentally
to some other use of this publication) without the written
permission of the copyright holder except in accordance with the
provisions of the Copyright, Designs and Patents Act 1988 or under
the terms of a licence issued by the Copyright Licensing Agency Ltd,
90 Tottenham Court Road, London, England W1T 4LP. Applications for
the copyright holder's written permission to reproduce any part
of this publication should be addressed to the publisher

Permissions may be sought directly from Elsevier's Science and Technology Rights
Department in Oxford, UK: phone: (+44) (0) 1865 843830; fax: (+44) (0) 1865
853333; e-mail: permissions@elsevier.co.uk. You may also complete your request
on-line via the Elsevier homepage (http://www.elsevier.com), by selecting
'Customer Support' and then 'Obtaining Permissions'

British Library Cataloguing in Publication Data
A catalogue record for this book is available from the British Library

Library of Congress Cataloguing in Publication Data
A catalogue record for this book is available from the Library of Congress

ISBN   0 7506 5106 7

For information on all Butterworth-Heinemann
publications visit our website at www/bh/com

Typeset by Replika Press Pvt Ltd, India
Printed and bound in Great Britain by Biddles Ltd, *www.biddles.co.uk*

Contents

Preface *xiii*
List of contributors *xv*

Part 1 Testing

1 Analysis of fresh concrete 1/3
Alan Williams

   1.1 Introduction   1/3
   1.2 British Standards covering fresh analysis   1/4
   1.3 Tests for cement content   1/4
         1.3.1 Calibration samples   1/4
         1.3.2 Test samples   1/4
         1.3.3 Applicability of test methods   1/5
         1.3.4 Buoyancy (old BS 1881) method   1/6
         1.3.5 Constant volume (RAM) method   1/8
         1.3.6 Pressure filter (Sandberg) method   1/12
   1.4 Tests for pfa content   1/14
         1.4.1 Calibration   1/15
         1.4.2 Determining the particle density   1/15
         1.4.3 PFA test   1/17
   1.5 Tests for ggbs content   1/17
         1.5.1 Chemical test apparatus   1/17
         1.5.2 Chemical test procedure   1/18
         1.5.3 Calibration   1/18
         1.5.4 GGBS testing   1/19
   1.6 Tests for water content   1/19
         1.6.1 High-temperature method   1/19

	1.6.2	Microwave oven method	1/20
	1.6.3	Oven-drying method	1/21
1.7	Aggregate grading	1/21
	1.7.1	Buoyancy method	1/21
	1.7.2	RAM method	1/21
	1.7.3	Pressure filter (Sandberg) method	1/22
1.8	Summary	1/22
Reference	1/22

2 Strength-testing machines for concrete — 2/1
J. B. Newman

2.1	Introduction	2/1
2.2	Uniaxial compression testing	2/1
	2.2.1 Introduction	2/1
2.3	Specification for compression testing machines	2/5
2.4	Verification procedures	2/6
	2.4.1 Force transfer	2/6
	2.4.2 Force calibration	2/7
	2.4.3 Comparative cubes	2/8
2.5	Tensile strength testing	2/8
2.6	Flexural strength testing	2/10
References		2/11

3 Accelerated strength testing — 3/1
Tony Binns

3.1	Aims of accelerated and early-age testing	3/1
3.2	Principles	3/2
3.3	British Standards procedures	3/3
	3.3.1 General procedures	3/3
	3.3.2 Test report – mandatory information	3/5
	3.3.3 Test report – optional information	3/5
3.4	American Society for Testing and Materials (ASTM) procedures	3/5
	3.4.1 General procedures	3/5
3.5	Other national standards	3/7
3.6	Other research	3/7
3.7	Applications of accelerated and early-age testing	3/8
	3.7.1 Control by prediction of 28-day strength	3/10
	3.7.2 Conformity	3/13
3.8	Conclusion	3/15
References		3/16

4 Analysis of hardened concrete and mortar — 4/1
John Lay

4.1	Aims and objectives	4/1
4.2	Brief history	4/1
4.3	Introduction	4/2

4.4	Reasons for analysis	4/2
4.5	Information that can be obtained by analysis	4/2
4.6	Sampling procedures	4/3
	4.6.1 General	4/3
	4.6.2 Sample types	4/4
	4.6.3 Number of samples	4/5
	4.6.4 Sample preparation	4/6
4.7	Determination of cement content of concrete	4/6
4.8	Analysis of mortar to determine mix proportions	4/9
4.9	Other determinations	4/11
	4.9.1 Determination of sulphate content	4/11
	4.9.2 Determination of chloride content	4/11
	4.9.3 Determination of alkalis content	4/11
	4.9.4 Determination of original water/cement ratio of concrete	4/11
	4.9.5 Aggregate grading	4/12
	4.9.6 GGBS content	4/12
	4.9.7 Carbon dioxide	4/13
	4.9.8 Admixtures	4/13
4.10	Accuracy and precision of determined cement content of concrete	4/13
4.11	Accuracy and precision of determined mix proportions of mortar	4/14
4.12	Summary	4/14
Acknowledgements	4/14
References	4/15
Further reading	4/15

5 Core sampling and testing · 5/1
Graham True

5.1	Introduction	5/1
5.2	The current situation regarding standards and guidance	5/1
5.3	Current core sampling, planning and interpretation procedures	5/2
	5.3.1 Reasons for taking and testing cores	5/2
	5.3.2 Planning and preliminary work before drilling cores	5/3
	5.3.3 Size, number of cores, location and drilling procedures	5/4
	5.3.4 Location and drilling of cores	5/5
	5.3.5 Visual examination and measurements	5/5
	5.3.6 Core preparation, conditioning and testing for density,	
	excess voidage and compressive strength	5/6
	5.3.7 Other tests	5/7
	5.3.8 Converting core strength to *in-situ* cube strength and	
	potential strength	5/8
	5.3.9 Interpretation of results and worked examples	5/10
5.4	Worked examples	5/12
	5.4.1 Example 1	5/12
	5.4.2 Example 2	5/14
5.5	Updating CSTR No. 11	5/15
	5.5.1 Obtaining the required new data	5/15
	5.5.2 Results of the new data	5/16
	5.5.3 Other considerations	5/17

	5.5.4 The effect of voidage and potential density on potential strength estimates	5/19
	5.5.5 Interpretation options	5/19
References		5/21

6 Diagnosis, inspection, testing and repair of reinforced concrete structures — 6/1
Michael Grantham

6.1	Introduction	6/1
6.2	What is concrete?	6/2
	6.2.1 Cement	6/2
	6.2.2 Water	6/2
	6.2.3 Aggregate	6/3
	6.2.4 Steel	6/3
	6.2.5 Admixtures	6/3
6.3	Recognizing concrete defects	6/4
	6.3.1 Structural failure	6/4
	6.3.2 Corrosion of steel	6/4
	6.3.3 Alkali–silica reaction	6/5
	6.3.4 Freeze–thaw damage	6/5
	6.3.5 Shrinkable aggregates	6/6
	6.3.6 Chemical attack	6/6
	6.3.7 Fire damage	6/9
	6.3.8 Poor-quality construction	6/10
	6.3.9 Plastic cracking	6/11
	6.3.10 Thermal cracking and delayed ettringite formation	6/11
6.4	Investigation of reinforced concrete deterioration	6/12
	6.4.1 The two-stage approach	6/12
	6.4.2 Visual survey	6/16
	6.4.3 Covermeter survey	6/16
	6.4.4 Ultrasonic pulse velocity measurement (PUNDIT)	6/17
	6.4.5 Chemical tests	6/27
	6.4.6 Depth of carbonation	6/29
	6.4.7 Compressive strength determination	6/31
	6.4.8 Petrographic examination	6/33
	6.4.9 Surface hardness methods	6/35
	6.4.10 Radar profiling	6/39
	6.4.11 Acoustic emission	6/41
	6.4.12 Infrared thermography	6/42
6.5	Testing for reinforcement corrosion	6/43
	6.5.1 Half cell potential testing	6/43
	6.5.2 Resistivity	6/46
	6.5.3 Corrosion rate	6/47
References		6/51

Part 2 Repair

7 Concrete repairs 7/3
Michael Grantham

7.1	Patch repairs	7/3
	7.1.1 Patch repairing carbonation-induced corrosion	7/3
	7.1.2 Patch repairing chloride-induced corrosion	7/5
	7.1.3 The incipient anode effect	7/5
7.2	Cathodic protection	7/6
	7.2.1 Basic electrochemistry	7/6
	7.2.2 Corrosion	7/6
	7.2.3 Reactivity	7/7
	7.2.4 Cathodic protection	7/7
	7.2.5 Practical anode systems	7/8
7.3	Electrochemical chloride extraction (desalination) and realkalization	7/9
	7.3.1 Introduction	7/9
	7.3.2 The mechanisms of corrosion of steel in concrete	7/9
	7.3.3 Electrochemical processes	7/10
	7.3.4 Chloride removal	7/11
	7.3.5 Realkalization	7/15
	7.3.6 Conclusion	7/18
7.4	Corrosion inhibitors	7/18
	Acknowledgements	7/19
	References	7/19

Part 3 Quality and standards

8 Quality concepts 8/3
Patrick Titman

8.1	Introduction	8/3
8.2	Definitions	8/3
8.3	Systems management standards	8/5
	8.3.1 The primacy of ISO 9001	8/5
	8.3.2 Understanding the ideas of ISO 9001: 2000	8/5
	8.3.3 Understanding the text of ISO 9001: 2001	8/8
	8.3.4 Reconciling ideas and text	8/10
	8.3.5 Procedures and method statements	8/13
	8.3.6 The family of systems management standards	8/14
8.4	Third-party registration and sector schemes	8/16
	8.4.1 Agrément Board schemes	8/17
	8.4.2 CE marks	8/17
	8.4.3 Sector schemes	8/18
8.5	Self-certification and quality control	8/20
8.6	Details of ISO 9001	8/22
	8.6.1 General	8/22
	8.6.2 System and product review in the management standards	8/22

	8.6.3	Purchasing, subcontract and materials control (ISO 9001: 2000: scn. 7.4)	8/23
	8.6.4	Nonconformity and improvement	8/25
	8.6.5	Audit and review	8/27
8.7	Laboratory management		8/29
	8.7.1	Introduction	8/29
	8.7.2	The ISO 17025 framework	8/29
	8.7.3	Management of concrete sampling and testing	8/30
References			8/30

9 Quality control   9/1
Lindon Sear

9.1	Introduction	9/1
9.2	Control charts	9/2
9.3	Shewhart charts	9/3
	9.3.1 Monitoring the mean strength	9/3
	9.3.2 Monitoring the standard deviation	9/5
	9.3.3 Further analysis for trends	9/7
	9.3.4 Conclusions	9/7
9.4	Cusum charts	9/8
	9.4.1 The history of the Cusum system of quality control	9/8
	9.4.2 Controlling the strength of concrete in practice	9/9
	9.4.3 Controlling the mean strength of the concrete	9/9
	9.4.4 Controlling the standard deviation of the concrete	9/11
	9.4.5 Monitoring the accuracy of the predicted 28-day strength from the early test results	9/12
	9.4.6 Properties of the Cusum system of quality control	9/13
	9.4.7 Cusum charts in practice	9/15
	9.4.8 The implications of taking action	9/16
9.5	Compliance or acceptance testing	9/17
9.6	Operating-Characteristic (O-C) curves	9/18
9.7	Producer's and consumer's risk	9/19
9.8	Experimental design	9/20
References		9/23
Further reading		9/24

10 Statistical analysis techniques in ACT   10/1
Stephen Hibberd

10.1	Introduction	10/1
10.2	Overview and objectives	10/2
10.3	Sample data and probability measures	10/2
	10.3.1 Random variation	10/2
	10.3.2 Sample data	10/3
	10.3.3 Representation of data	10/3
	10.3.4 Quantitative measures	10/5
	10.3.5 Population values	10/6
	10.3.6 Probability	10/6
	10.3.7 Probability functions	10/7

	10.3.8 Expected values	10/8
	10.3.9 Normal distribution	10/8
	10.3.10 Calculation of probability values (from standard tables)	10/9
	10.3.11 Standardized normal variate	10/10
	10.3.12 Example	10/11
	10.3.13 Critical values	10/12
10.4	Sampling and estimation	10/13
	10.4.1 Sample statistics	10/13
	10.4.2 Large-sample statistics (normal distribution)	10/13
	10.4.3 Small-sample statistics (t-distribution)	10/14
	10.4.4 Confidence intervals	10/16
	10.4.5 Control charts	10/17
	10.4.6 Comparison of means	10/18
	10.4.7 Comparison of variances	10/20
10.5	Significance tests	10/21
	10.5.1 Hypothesis testing	10/21
	10.5.2 Comparison of means	10/24
	10.5.3 Comparison of variances	10/25
	10.5.4 Significance and errors	10/25
10.6	Regression models	10/26
	10.6.1 Correlation	10/26
	10.6.2 Regression – least squares method	10/28
	10.6.3 Correlation coefficient	10/29
	10.6.4 Example – Case 2 Beam deflection	10/31
	10.6.5 Analysis of residuals	10/31
	10.6.6 Extension to multivariate	10/33
	10.6.7 Fit of regression curves and confidence lines	10/33
10.7	Statistical formulae and tables	10/33
Further reading		10/40

11 Standards, specifications and codes of practice 11/1
Bryan K. Marsh

11.1	Aims and objectives	11/1
11.2	Introduction	11/1
11.3	Standards, specifications and codes of practice	11/2
	11.3.1 Standards	11/2
	11.3.2 Codes of practice	11/2
	11.3.3 Specifications	11/2
	11.3.4 Other relevant documents	11/3
	11.3.5 The role and status of standards, standard specifications and codes of practice	11/3
	11.3.6 Selection of appropriate standards and codes of practice	11/4
	11.3.7 Worldwide use of standards	11/5
	11.3.8 European Standards and International Standards	11/5
11.4	Prescription-based standards and performance-based standards	11/8
	11.4.1 The prescription-based approach	11/8
	11.4.2 The performance-based approach	11/9
11.5	The treatment of durability in standards, codes of practice and standard specifications	11/10

11.5.1	Introduction	11/10
11.5.2	Design for durability	11/10
11.5.3	Constituent materials	11/11
11.5.4	Constituent materials test methods	11/14
11.5.5	Classification of exposure environment	11/14
11.5.6	Resistance to degradation	11/18
11.5.7	Specification, production and conformity	11/24
11.5.8	Fresh concrete test methods	11/25
11.5.9	Hardened concrete test methods	11/25
11.5.10	Execution, or workmanship	11/25
11.5.11	Concrete products	11/26
11.5.12	Some limitations of standards	11/26
11.6	Specifications	11/27
11.6.1	Introduction	11/27
11.6.2	Typical specification clauses	11/28
11.6.3	Writing specifications	11/29
11.6.4	Using specifications	11/29
References		11/29
Further reading		11/31
Index		I/1

Preface

The book is based on the syllabus and learning objectives devised by the Institute of Concrete Technology for the Advanced Concrete Technology (ACT) course. The first ACT course was held in 1968 at the Fulmer Grange Training Centre of the Cement and Concrete Association (now the British Cement Association). Following a re-organization of the BCA the course was presented at Imperial College London from 1982 to 1986 and at Nottingham University from 1996 to 2002. With advances in computer-based communications technology the traditional residential course has now been replaced in the UK by a web-based distance learning version to focus more on self-learning rather than teaching and to allow better access for participants outside the UK. This book, as well as being a reference document in its own right, provides the core material for the new ACT course and is divided into four volumes covering the following general areas:

- constituent materials
- properties and performance of concrete
- types of concrete and the associated processes, plant and techniques for its use in construction
- testing and quality control processes.

The aim is to provide readers with an in-depth knowledge of a wide variety of topics within the field of concrete technology at an advanced level. To this end, the chapters are written by acknowledged specialists in their fields.

The book has taken a relatively long time to assemble in view of the many authors so the contents are a snapshot of the world of concrete within this timescale. It is hoped that the book will be revised at regular intervals to reflect changes in materials, techniques and standards.

John Newman
Ban Seng Choo

List of contributors

Tony Binns Training Workshops,
PO Box 5328, Slough SL2 3FL, UK

Michael Grantham
MG Associates Construction Consultancy Ltd, 11 The Quadrant, Manor Park Crescent, Edgeware, Middlesex HA8 7LU, UK

Stephen Hibberd
School of Mathematical Sciences, The University of Nottingham, University Park, Nottingham NG7 2RD, UK

John Lay
RMC Aggregates (United Kingdom) Ltd, RMC House, Church Lane, Bromsgrove B61 8RA, UK

Bryan K. Marsh
Arup Research and Development,
13 Fitzroy Street, London W1T 4BQ, UK

John Newman
Department of Civil Engineering, Imperial College, London SW7 2BU, UK

Lindon Sear
UK Quality Ash Association, Regent House, Bath Avenue, Wolverhampton, West Midlands WV1 4EG, UK

Patrick Titman
'Cold Harbour', Cold Harbour Lane, Marlborough, Wiltshire SN8 1BJ

Graham True
GFT Materials Consultancy, 19 Hazel Road, Dudley DY1 3EN, UK

Alan Williams
Wexham Developments, Unit 8, Young's Industrial Estate, Paices Hill, Aldermaston, Reading RG7 4PW, UK

PART 1
Testing

1

Analysis of fresh concrete

Alan Williams

1.1 Introduction

The analysis of fresh concrete comprises a range of on-site tests which can be carried out to determine the cement, pulverized-fuel ash (pfa), ground granulated blast furnace slag (ggbs), and water content of the original concrete mix, and the aggregate grading. Very fine additions to the concrete such as microsilica are treated as part of the fines content of the mix.

The main advantage of fresh analysis is that it gives the concrete technologist a set of tests which can be performed on-site, as the concrete is being placed. As part of a quality control scheme, testing at regular intervals will provide a guide to the variability of the concrete as supplied to, or mixed, on-site. Using a rapid analysis machine (RAM), the operator is able to determine the cement content within 15 minutes of taking the samples. The times taken for other test methods are shown in Table 1.2.

The main disadvantage of fresh analysis is that all the tests are based on the physical separation of the cement, pfa and ggbs from the other constituents of the concrete, and an allowance has to be made for the fines or silt content of the aggregate. Calibration procedures using site materials and regular checks on silt content of the aggregate can reduce this problem to an acceptable level. The initial cost of the equipment would prohibit its use on small sites, but Clear (1988) has shown that if the concrete samples are stored below the temperature at which hydration ceases, then they can be transported in a cold box to a test house for analysis using any of the techniques described in this chapter.

1.2 British Standards covering fresh analysis

Early versions of BS 1881: Part 2: 1970 had a single method, the Buoyancy Method, for determining the cement content of fresh concrete. The current Standard BS 1881: Part 128: 1997, covers three methods for cement content and has additional tests for pfa, slag, water content and aggregate grading, all of which are described in detail in later sections.

1.3 Tests for cement content

For all methods of analysis the tests for cement content involve the same four basic steps:

1. calibration of the test method using site materials prior to testing site concrete
2. collecting representative samples of the fresh concrete for testing
3. the physical separation of the cement-sized particles from the remainder of the concrete sample using the chosen test method
4. determination of the cement content using the previously established calibration

If there are cement replacement materials in the mix (pfa or ggbs) then there is an additional step to determine the content of the replacement material as described in sections 1.4 and 1.5 below.

If the concrete is air-entrained, then a chemical is stirred into the sample prior to testing, to remove the air.

1.3.1 Calibration samples

It is important that the test method chosen is calibrated using representative samples of the site materials. If the concrete comes from a readymix supplier, then representative samples should be obtained from the supplier. This is particularly important when establishing the fines or silt (<150 micron) content of the aggregates. From the concrete mix design, calculate the saturated surface dry (SSD) mass of the coarse aggregates and the fine aggregates required to make a test sample of the required size.

Each bulk sample of aggregate must be reduced to provide a subsample containing sufficient aggregate for the individual analysis.

1.3.2 Test samples

All of the analysis methods require the testing of duplicate samples, as the difference in value between the duplicates is a good guide to the accuracy of the sampling and testing method. For cement testing, a variation greater than 20 kg/m^3 between the duplicate samples would invalidate the test (BS 1881: Part 128: A.7 Repeatability).

In the old BS 1881 method, the sample reduction method was advocated for obtaining test samples of the requisite size. This entailed reducing a bulk sample by successive coning and quartering until the required sample size was attained.

Coning and quartering consists of obtaining a fresh concrete sample of the required size and heaping it up to form an inverted cone on a steel plate. The concrete is then moved with a shovel to form a second inverted cone on a clean part of the plate, which helps to inter-mix the sample. This process is repeated to form a new cone, again on a clean part of the steel plate. Finally the shovel is inserted vertically into the centre of the cone and the concrete spread out to form an approximate circle of even depth. This circle is separated into quarters and each opposite pair of quarters combined to form two test samples.

Testing samples obtained in this way result in a low value of cement content as some of the finer cement-sized particles are lost in the reduction process. Experience has shown that a more effective method is to sample the required amount directly into plastic buckets.

Sampling from mixing or agitating trucks

To sample from a truck mixer, allow the first 0.5 m^3 of concrete to be discharged and pre-coat a metal scoop with cement and fines by holding it in the discharge stream. Then take scoopfuls of concrete from the moving stream to provide duplicate samples of the mass, shown in Table 1.1, placing the scoopfuls alternately in two clean pre-weighed plastic buckets. Do not sample from the last 0.5 m^3 of concrete to be discharged.

Sampling from bulk quantities of concrete

Again pre-coat the scoop then collect in a clean plastic bucket the required number of scoopfuls of concrete to provide a sample of appropriate mass shown in Table 1.1.

Table 1.1 Sample mass required for testing

Analysis method	Concrete sample mass (kg)	Test sample mass (kg)
Buoyancy (BS 1881)	8 ± 1	4 ± 0.5
Constant Volume	16 ± 2	8 ± 1
Pressure Filter (Sandberg)	6 ± 1	3 ± 0.5
Water Content	5 ± 1	2.5 ± 0.5

1.3.3 Applicability of test methods

The standard procedure for the Buoyancy method does not provide a sample of the material passing the 150-micron sieve for further analysis if required (Table 1.2).

Table 1.2 Applicability of test methods

Analysis method	Cementitious content	Aggregate content	Water content	Availability of cementitious material	Test duration (min)
Buoyancy (BS 1881)	Yes	Yes	*	No	90
Constant volume	Yes	Yes	*	Yes	15
Pressure filter (Sandberg)	Yes	Yes	*	Yes	90
Water Content	No	No	Yes	No	30

* Water content can be calculated by difference, but the accuracy of the result will reflect any errors in the determination of the cementitious and aggregate contents.

1.3.4 Buoyancy (old BS 1881) method

In this method the test sample is weighed in air, then in water, then washed over a nest of sieves to separate the cement and fines. The washed aggregate is weighed in water and the proportions of cement, coarse aggregate, fine aggregate and water calculated on the basis of predetermined values of relative densities (Figure 1.1).

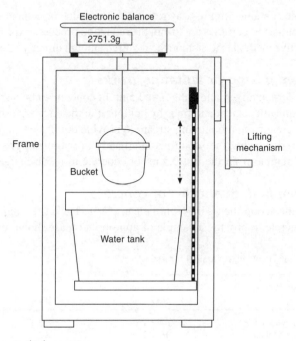

Figure 1.1 Buoyancy method apparatus.

Calibration
Relative densities

Take sub-samples of the fine aggregate fractions, allowing for the free water in the aggregate, to provide a fine coarse aggregate sample of the required mass, and place into one of the clean round-bottomed containers. Repeat for the coarse aggregate, and place this sample onto a 5-mm sieve and wash with a spray of water for 2 minutes to remove particles finer than the 5 mm through the sieve, into the container with the fine aggregate.

Transfer the washed coarse aggregate into a clean round-bottomed container and fill with water to within 25 mm of the lip. Stir for about 1 minute to remove any entrapped air, then immerse the container in the water tank and weigh. Repeat the process with the container holding the fine aggregate.

If the aggregates have been oven-dried, water absorption will occur and the mass shown on the balance will change as the water is absorbed. If this is the case, stir and reweigh the container and contents at 10-minute intervals until the change in mass is less than 0.5 g. Record the time taken to reach this condition. Record the final mass in water as B_a for the coarse aggregate, and B_s for the fine aggregate.

Carefully drain the water from each container, and dry the coarse and fine aggregates separately to the saturated surface dry condition in accordance with BS: 812: Part 2. Record mass in air as A_a for the coarse aggregate (SSD) and A_s for the fine aggregate (SSD). Calculate the relative densities:

$$\text{Relative density of coarse aggregate} = \frac{A_a}{A_a - B_a}$$

$$\text{Relative density of fine aggregate} = \frac{A_s}{A_s - B_s}$$

where
A_a = mass in air of the coarse aggregate (SSD)
A_s = mass in air of the fine aggregate (SSD)
B_a = mass in water of the coarse aggregate
B_s = mass in water of the fine aggregate

Repeat the above operations three more times with new sub-samples then calculate the average values, save the fine aggregate samples for determining the fines correction factor.

Determination of fines correction factor

Place one of the fine aggregate samples on a 150-micron sieve and wash under a spray of water for about 10 minutes. Wash the aggregate retained on the sieve into a clean round-bottomed container and determine its weight in water as described previously. Calculate the fines correction factor C_s from

$$C_s = \frac{B_s}{D_s}$$

Repeat on the other three samples to obtain the mean value.

Analysis of concrete

Place one of the duplicate test samples in a clean, dry round-bottomed container and determine its mass in air W, then immerse and re-weigh to obtain the sample mass in water w. If the water in the tank becomes contaminated, change it to prevent a change in its density. Transfer the sample to the nest of sieves, 5 mm over the 150 micron, and wash the concrete until it is free from cement (for at least 2 minutes).

Transfer the clean coarse aggregate from the 5-mm sieve to a clean container and immerse the container in the water and determine the mass in water of the coarse aggregate, w_a. Wash the fine aggregate through the 150-micron sieve for a further 10 minutes then transfer to a container, immerse and determine w_s. Repeat the above procedure with the second duplicate test sample.

Calculation of mass of each constituent

Calculate the mass of each constituent:

(a) the mass of the coarse aggregate W_a

$$W_a = w_a \times F_a$$

(b) the mass of fine aggregate W_s

$$W_s = w_s \times F_s \times C_s$$

Analysis of fresh concrete

(c) the mass of cement W_c

$$W_c = (w - w_a - (w_s \times C_s)) \times F_c$$

where

$F_a = \dfrac{\text{relative density}}{\text{relative density} - 1}$ for the coarse aggregate

$F_s = \dfrac{\text{relative density}}{\text{relative density} - 1}$ for the fine aggregate

$F_c = \dfrac{\text{relative density}}{\text{relative density} - 1}$ for the cement

C_s is the fines correction factor
W is the mass of the concrete sample in air
w is the mass of the concrete sample in water
w_a is the mass of coarse aggregate in water
w_s is the mass of the fine aggregate in water

The mass of each constituent per cubic metre of concrete (in kg/m^3)

$$= \dfrac{\text{mass of constituent}}{\text{mass of test sample}} \times \text{mass}/m^3 \text{ of compacted fresh concrete}$$

If the difference between the duplicate test samples exceeds 20 kg/m^3, then BS 1881: Part 128 requires you to discard the test results.

1.3.5 Constant volume (RAM) method

In this analysis method the concrete test sample is weighed and transferred to the elutriation (flow separation) column of a rapid analysis machine (RAM). The machine separates the fine cement-sized particles from the concrete and 10 per cent of the resultant suspension is diverted through a vibrating 150-micron sieve into a conditioning vessel in which the suspension is flocculated. All of the cement-sized particles come out of suspension and settle in a removable vessel (constant-volume vessel) attached to the bottom of the conditioning vessel. Excess water is removed by siphons until the level drops to a fixed point at which the siphon breaks, leaving a constant volume of water and flocculated material in the vessel. The constant volume vessel is removed and weighed, and the mass of cement and fine sand in the total suspension is determined by reference to a calibration chart.

Operation of the RAM machine
The basic RAM test procedure consists of:

1 collecting two 8 kg concrete samples in plastic buckets and weighed accurately;
2 PRIMING the RAM machine then washing all of the sample from the bucket into the machine via the loading hopper;
3 starting the automatic cycle;

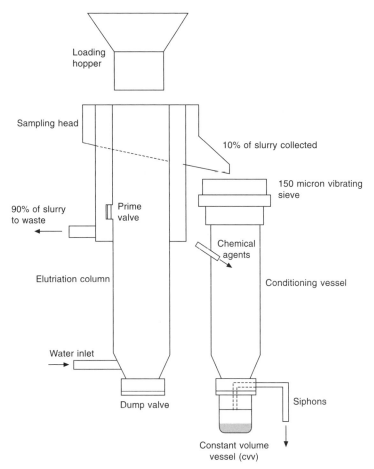

Figure 1.2 Rapid analysis machine (RAM).

4 after the buzzer sounds, removing the constant volume vessel from the machine, and weighing the vessel and contents on a 2 kg balance, and recording the mass as W_{cs};
5 recovering the aggregate from the elutriation column and from the sieve;
6 washing clean the inside of the machine and the sieve ready for the next test;
7 repeating for the second sample.

Calibration

The RAM Machine is calibrated using 'prepared' aggregate and known volumes of cement.

(a) *'Prepared' aggregate*   From the aggregate sub-samples weigh out an 8 kg sample allowing for the moisture content of the aggregate. Place this sample in a bucket, weigh and then add an equivalent volume of water and stir with a metal rod. Carry out a RAM test as described previously and when the machine has finished its cycle, recover the clean 'prepared' aggregate from the elutriation column into a clean plastic bucket. By putting the aggregate through a test cycle, all of the cement-sized particles are removed.

(b) *Test with known cement values*   Using the 'prepared' aggregate, carry out five RAM

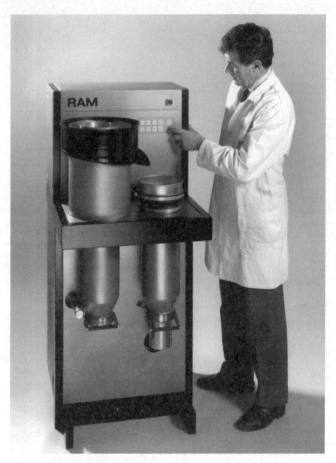

Figure 1.3 Rapid analysis machine (RAM).

tests with zero cement content, recovering the aggregate for re-use each time. If the range of the five recorded values is greater than 2.0 g, discard and repeat the tests. If the range is acceptable, then record the average of the five readings as W_0.

Add 750 g of cement to the 'prepared' aggregate in the bucket, and carry out a RAM test. Repeat to get a further four values with 750 g cement content. If the range of the five recorded values is greater than 3.5 g, discard the results and carry out five more tests. If the range is acceptable, then record the average of the five readings as W_{750}.

Repeat the tests with 1500 g of cement to obtain W_{1500}, with an allowable range of 5.0 g on the constant volume mass.

Draw a graph of constant volume weight against cement content, the calibration line is the straight line joining the points corresponding to the 750 g cement value and the 1500 g cement value (Figure 1.4).

(c) *Determination of the fines correction value* This procedure is similar to the previous tests except that representative aggregate containing cement-sized particles is used in place of the 'prepared' aggregate in the test sample. Make up a test sample containing 1000 g of cement and proportional amounts of fine and coarse aggregate in accordance

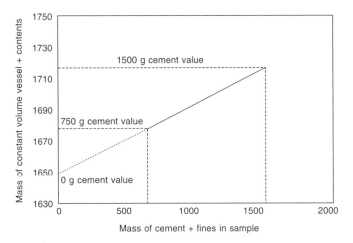

Figure 1.4 RAM calibration graph.

with the mix design. Weigh the test sample and carry out a RAM test exactly as before. At the end of the test, weigh the constant volume vessel and determine the apparent cement content from the calibration line. Determine the fines content by difference:

fines content (g) = indicated cement content (g) − 1000 g

then

$$\text{fines content kg/m}^3 = \frac{\text{fines content (g)} \times \text{mass/m}^3}{\text{mass of sample (g)}}$$

Repeat to get five values, the average of the five values is the fines or silt correction value.

Analysis of concrete

1. take duplicate test samples of approximate weight 8 kg;
2. weigh the first test sample and record the mass W (*note*: if air-entrained concrete, add extra water to the sample followed by 10 ml of tri-n-butyl phosphate, stir thoroughly for 2 minutes to remove the air);
3. test the sample in the RAM machine as described above and record the weight W_{cvv};
4. read off apparent cement content W_{ac} by reference to the cement calibration line;
5. then

$$\text{apparent cement content kg/m}^3 = \frac{W_{ac}}{W} \text{ mass/m}^3 \text{ of fresh concrete}$$

6. repeat for the duplicate test sample.

Then

Cement content = apparent cement content − fines correction value (all in kg/m^3)

If the difference in the two results is greater than 20 kg/m^3, then the sampling technique is suspect, and the results should be discarded.

1.3.6 Pressure filter (Sandberg) method

In this method the 3 kg concrete sample is weighed, then placed in a bottle and agitated with water to start separating the cement-sized particles from the rest of the concrete mix. The mixture is then washed over a nest of sieves to separate the cement and fines passing a 150-micron sieve. The washings are filtered using a pressure filter and the solids are dried and weighed. Allowances are made for the fines mixed with the cement and the solubility of the cement (see Figure 1.5).

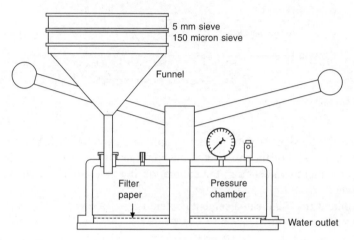

Figure 1.5 Pressure filter.

Calibration
Determination of the fines content of the aggregate

From the concrete mix design calculate the saturated surface dry (SSD) mass of the coarse aggregates and the fine aggregates required to make a test sample of 3 kg. Place the required amounts of aggregate in trays and dry to constant mass. Check the aggregate fine to coarse ratio and adjust if necessary. Transfer the weighed batch of aggregate to the mixing bottle and add 2 l of water.

Seal the bottle and shake vigorously, preferably on a bottle roller, to ensure the separation of any passing 150-micron particles from the aggregate.

Pour the liquid contents of the bottle, together with any fines, through a nest of sieves with the 5-mm sieve at the top and the 150-micron sieve at the bottom, including any intermediate sieves that are considered necessary to protect the 150-micron sieve.

Continue rinsing the aggregate in the bottle and on the sieves until the wash water is free from fine material. Then transfer the aggregate remaining in the bottle to the top sieve and allow to drain for 2 to 3 minutes.

After draining, transfer the aggregate to a tray and dry to constant mass. Grade the aggregate over a 5 mm sieve, any intermediate sieve(s), and a 150-micron sieve to complete the separation of the passing 150-micron aggregate. Weigh the aggregate larger than 150 micron.

Calculate the percentage fines content of the aggregate(s):

$$W_{s1} = \frac{W_1 - W_2}{W_2} \times 100$$

where
W_{s1} is the percentage of fines content; for convenience this is based on the dry, washed aggregate
W_1 is the mass of dry aggregate before washing
W_2 is the mass of dry aggregate after washing.

Determination of the cement correction factor

Place a 1 kg sample of cement in a tray and dry in an oven at $200 \pm 5°C$ for 60 ± 10 minutes. Weigh a test sample of the cement W_3 equal to the amount expected in a concrete test sample. Prepare a slurry of the test sample of cement with half of its mass of water by mixing for 4 minutes in a suitable container; allow the slurry to stand for 30 ± 5 minutes. Place a weighed filter paper, dried for not more than 1 hour, in the pressure filter and assemble the filter. Insert the neck of the charging funnel into the top opening of the pressure filter, and support the 150-micron sieve on the funnel. Wash the cement slurry through the 150-micron sieve into the pressure filter, continue washing until the wash water becomes clear. Remove the funnel, seal the pressure filter, and apply a pressure of 0.20 MN/m^2 until filtration is complete.

Dismantle the pressure filter and carefully transfer the paper and retained cement, on a tray, to the oven. Dry at $200 \pm 5°C$ for 60 ± 10 minutes. Weigh the filter paper and cement, deduct the mass of the paper, and record the mass of recovered cement as W_4.

Calculate the percentage cement correction factor

$$W_{t1} = \frac{W_3 - W_4}{W_3} \times 100$$

where
W_{t1} is the percentage cement correction factor
W_3 is the mass of dry cement test sample
W_4 is the mass of dry cement retained on the filter paper.

Determination of the water absorption of the coarse and fine aggregates

Determine the water absorption of the coarse and fine aggregates in accordance with Paragraph 5 of BS 812: Part 2.

Analysis of concrete

Take duplicate test samples of 3 ± 0.5 kg. Weigh the first test sample W_s and immediately transfer it to the bottle. Add 2 litres of water, seal the bottle and agitate on the mechanical shaker for 10 to 20 minutes. Sieve, pressure filter and dry the filtrate exactly as described above. Weigh the filter paper and solids, deduct the mass of the filter paper, and record the mass of the solids as W_6.

Determination of the mass of coarse and fine aggregates

Transfer the whole of the aggregate on the sieves to drying pans or trays. Dry the aggregate to constant mass.

Grade and weigh the recovered dry aggregate and increase the masses to include the appropriate corrections for the absorption of the aggregates, and record the corrected mass of the coarse aggregate as W_7; the corrected mass of the fine aggregate as W_8; and the corrected mass of the dried aggregate that passes a 150-micron sieve as W_9 by assuming that the water absorption of this fraction is the same as for the fine aggregate. Calculate the mass of each constituent:

Mass of coarse aggregate in the test sample = W_t

Mass of fine aggregate in the test sample

$$= W_8 + \frac{W_s}{100} \times (W_7 + W_8)$$

Mass of cement in the test sample

$$= \left[W_6 + W_9 - \frac{W_s}{100}(W_7 + W_8) \right] \times \frac{100}{100 - W_t}$$

where
W_5 is the mass of the concrete test sample (in g)
W_6 is the mass of solids retained on the filter paper (in g)
W_7 is the mass of the recovered coarse aggregate (in g)
W_8 is the mass of the recovered fine aggregate (in g)
W_9 is the mass of recovered fine aggregate passing a 150-micron sieve (in g)
W_s is the percentage fines content of the combined aggregates
W_t is the percentage cement correction factor.

Determination of mass per cubic metre of fresh concrete
On a separate sample, taken at the same time and by the same procedure as the sample for analysis, determine the mass per cubic metre of the fully compacted fresh concrete using the method described in BS 1881: Part 107.

Calculation of mass of each constituent per cubic metre of concrete
The mass of each constituent per cubic metre of concrete (in kg/m^3)

$$= \frac{\text{mass of constituent}}{\text{mass of sample}} \times \text{mass/m}^3 \text{ of compacted fresh concrete}$$

If the difference between the two masses of cement per cubic metre on the duplicate test samples exceeds 20 kg/m^3, discard the test results.

1.4 Tests for pfa content

The pfa content of the fresh concrete mix can be determined by a chemical method, as in Appendix of BS 6610, Pozzolanic cement with pulverized-fuel ash as pozzolan, or the Particle Density Method described below. Before the pfa content can be found the total cementitious content of the concrete has to be determined by one of the analysis methods described previously. The cementitious material recovered is then weighed dry and in water to determine the particle density. As the material obtained from the cement tests can be either:

dry – pressure filter method
wet – RAM method

The sequence of the two weighings depends on which of these two analysis methods was used.

1.4.1 Calibration

From the concrete mix proportions, calculate the saturated surface dry (SSD) mass of the coarse aggregates and the fine aggregates which would be required to make a test sample containing 1000 g of cement. Make 10 samples, weigh batching the aggregate for each sample directly into a bucket.

Add 1000 g of cement to five of the samples and the same weight comprising 20 per cent of cement and 80 per cent pfa to each of the remaining five samples. Test for cement content using one of the methods previously described, then determine the particle density of the recovered cementitious material.

Calculate the mean and range of each set of five results at both the 0 per cent pfa and 80 per cent pfa content. The pfa calibration is the line between the 0 and 80 per cent pfa content providing the range of each set of results does not exceed 0.12 kg/m^3 as specified in BS 1881: Part 128.

1.4.2 Determining the particle density

The particle density is calculated from the dry and wet weights of the material recovered from the cement content test.

Wet weight of material
Fill a gas jar (Figure 1.6) with water until a convex meniscus forms at the brim. Slide the cover plate across the top ensuring no air bubbles are trapped. Wipe the external surfaces with the absorbent cloths and weigh and record its mass, then discard the water.

Pressure filter test
Transfer the dry cementitious material from the filter paper into the gas jar and continue as described below.

RAM test
Use a fine jet washer bottle to wash all the slurry from the RAM constant-volume vessel into the gas jar. After waiting at least one minute for the slurry to partially settle in the jar gently top up the gas jar with water to a point where a convex meniscus forms at the brim. Slide the gas jar cover plate across the top of the gas jar ensuring no air bubbles are trapped. Dry the outside and weigh the gas jar, cover plate and contents to the nearest 0.1 g.

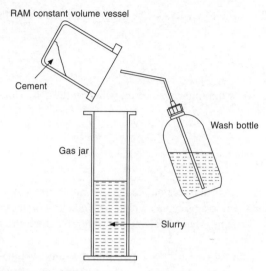

Figure 1.6 Recovery of cement into gas jar.

Dry weight of material
Pressure filter test
The dry weight of the material is determined as part of the Pressure Filter test (W_9) (see above).

RAM test
Weigh a glass oven tray and glass cloth to the nearest 0.1 g. Spread the cloth over the tray and wash the contents of the gas jar onto the cloth using a fine jet wash bottle. Fold the loose ends of the cloth to cover the sample and remove excess clear water by carefully sucking it into a wide jet wash bottle. Place the oven tray in a microwave oven of at least 1400 watt capacity, and dry for six minutes at full power. Remove the tray and contents from the oven, weigh to the nearest 0.1 g. Continue drying in 1-minute intervals until the reduction in mass is less than 0.3 g, about 9 minutes total drying time is required for calibration samples containing 1000 g of cement. Subtract the mass of the oven tray and cloth to obtain the mass of the dried material.

Calculate the particle density

$$PD = \frac{M}{M + M_0 - M_1}$$

where
PD is the particle density of the solids in the slurry in kg/litre
M is the mass of the dried slurry
M_0 is the mass of the water-filled gas jar and cover plate
M_1 is the mass of the gas jar, water, slurry and cover plate.

Figure 1.7 shows a typical calibration graph obtained using this method.

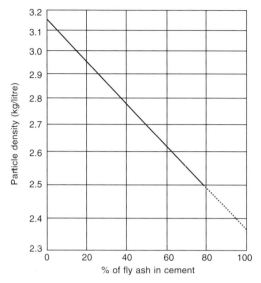

Figure 1.7 Particle density graph.

1.4.3 PFA test

Test duplicate samples for cementitous content, then for particle density and determine the pfa content of the concrete from the calibration line. Where the difference between the pfa content results for both samples is less than 5 per cent the test is valid and the mean of the two results is the measured pfa content.

1.5 Tests for ggbs content

Before the ggbs content can be determined the total cement content of the concrete is found using one of the methods described previously. A chemical test is then carried out on the dried cementitious material recovered to determine the sulphide content and hence the ggbs content. The process for recovering and drying the cementitious material is exactly the same as for pfa.

After drying, the material is ground for 15 seconds in a coffee grinder to homogenize it, then stored in an airtight container until tested. The dried residue can be stored for seven days without any significant effect on the results obtained. After 28 days' storage, the sample oxidizes and can produce erroneous results.

1.5.1 Chemical test apparatus

The apparatus consists of a glass reaction vessel fitted with a stopper, a vacuum pump and Drager Indicator Tubes for measuring hydrogen sulphide gas (Figure 1.8).

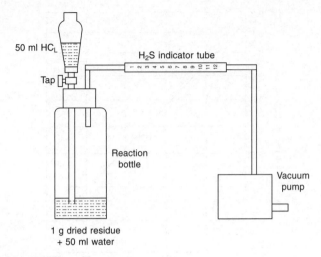

Figure 1.8 Slag test apparatus.

1.5.2 Chemical test procedure

1. measure 50 ml of tap water into the reaction vessel;
2. measure 50 ml of hydrochloric acid into the 50 ml cylinder;
3. using small pliers, break off the sealed ends of the indicator tube and place the outflow end (the end to which the arrow points) into the plastic tube that goes to the suction side of the pump;
4. connect the other end of the indicator tube to the reaction vessel-outflow tube;
5. turn on the pump;
6. weigh out a 1 g sample, noting the weight to 0.01 g;
7. place the sample in the reaction vessel and gently agitate to ensure even dispersion;
8. fit the stopper assembly into the reaction vessel ensuring the end of the inflow pipe is below the surface of the water;
9. open the valve allowing the acid to enter the reaction vessel;
10. gently agitate the reaction vessel for two minutes;
11. leave until a constant length of discoloration (to 1 mm) is attained (between 15 and 25 minutes for residue samples at 20 to 80 per cent ggbs respectively).

The corrected indicator tube reading is calculated:

$$R = \frac{T}{F}$$

where
R is the corrected tube reading (mm)
T is the measured tube reading (mm)
F is the manufacturers' calibration factor for the batch of tubes.

1.5.3 Calibration

Calibration is similar to that used for pfa except in this case:

five samples with 20 per cent ggbs/80 per cent cement
and five samples with 80 per cent ggbs/20 per cent cement

are tested and where the range of discolorations do not exceed 3.3 mm the results are acceptable. A typical calibration graph is shown in Figure 1.9.

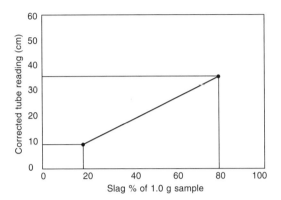

Figure 1.9 Slag test calibration graph.

1.5.4 GGBS testing

Test duplicate samples for cementitious content, then for sulphide content and determine the ggbs content of the concrete from the calibration line. Where the difference between duplicate samples is less than 5 per cent the test is valid and the mean of the two results is the measured pfa content.

1.6 Tests for water content

The determination of the water/cement ratio of concrete is normally based on the 'free' water content of the concrete at the time of mixing, where:

'Free' water content = total water content less the water absorption of the fine and coarse aggregates

1.6.1 High-temperature method

The total water content of fresh concrete is determined by drying duplicate test samples of 2.5 ± 0.5 kg to constant mass over a radiant heater or hot-plate. Weigh the test sample W_1, place it in the tray and heat it, taking care to ensure that the aggregate does not reach a temperature where spitting or decomposition occurs. During heating stir the test sample with a spatula to avoid local overheating. After 20 minutes, weigh the test sample and record the mass. Continue heating and re-weighing at 5-minute intervals until the difference between consecutive weighings is less than 0.1 per cent of the last recorded mass. Record the final dry mass as W_2:

$$\text{Total water content} = \frac{W_1 - W_2}{W_1} \times \text{mass/m}^3 \text{ of fresh concrete}$$

where
W_1 is mass of test sample
W_2 is mass of dried sample

Absorbed water content

Calculate the dried to constant weight masses per m³ of the coarse aggregate W_A and the fine aggregate W_F on the basis of the nominal mix proportions, then

$$\text{Absorbed water content} = (W_A \times F_A) + (W_F \times F_F) \text{ kg/m}^3$$

where
F_A = water absorption (% of dry mass) of coarse aggregate
F_F = water absorption (% of dry mass) of fine aggregate

Free water content

Free water content (kg/m³) = total water content − absorbed water content

If the difference between the two results is less than 10 kg/m³, the mean of the results is the free water content (kg/m³) of the sample.

1.6.2 Microwave oven method

The total water content of fresh concrete is determined by drying duplicate samples to constant mass in a microwave oven. The samples are wrapped in glass cloths and placed in oven-proof glass trays.

Drying method

Weigh and record the mass of the glass oven tray and a glass cloth, to the nearest 0.1 g. Spread the cloth over the tray so that there is an even lap over the four sides. Repeat for a second tray and cloth for the duplicate sample. Collect a test sample mass between 4 and 6 kg, thoroughly mix then divide into two equal halves, recovering all the fines and water, and place on the glass cloths in the two oven trays. Fold the loose ends of wrapper over to cover the samples, and place the first tray with its contents in the microwave oven. Dry the residue in the oven for six minutes at full power. Remove the tray and contents from the oven, weigh and record the mass to the nearest 0.1 g. As the oven tray is hot the balance should be protected with insulation.

Continue drying in 1-minute intervals and reweighing until the reduction in mass is less than 1.0 g. The total drying time of concrete samples is normally less than 20 minutes. Repeat the exercise for the second sample.

Weigh and record the mass W_2 of each dry sample, cloth and tray. Calculate the mass of water lost from each sample by subtracting the dry mass of each sample, cloth and tray from the mass of wet sample, cloth and tray.

$$\text{Total water content (kg/m}^3\text{)} = \frac{W_1 - W_2}{W_1} \times \text{mass/m}^3 \text{ of compacted fresh concrete}$$

where
W_1 is mass of test sample
W_2 is mass of dried sample

The absorbed water content and free water content are calculated as in the previous section.

1.6.3 Oven-drying method

This method is similar to the previous two except the duplicate samples are weighed then oven-dried at a temperature of 200°C for 16 hours.

1.7 Aggregate grading

Because the tests for cement content involve the physical separation of the cement from the aggregate. It is possible to recover the aggregate, using the following methods.

1.7.1 Buoyancy method

Recover the aggregate from the 5-mm and 150-micron sieves and sieve to obtain a grading and determine the mass of particles passing the 150-micron sieve using the fines correction factor C_s. Combining these two sets of figures gives the aggregate grading.

1.7.2 RAM method

In the RAM test, the aggregate has to be recovered from three sources.

The elutriation column
By opening the dump valve, the aggregate from the elutriation column can be collected and sieved to obtain a grading.

The sieve
Recover the sand retained on the 150-micron sieve and determine its grading. This grading represents approximately 10 per cent of the sand carried away by the water during the test. The total mass of sand carried away in the waste slurry using the RAM is calculated using the RAM sampling factor:

$$F_{sf} = \frac{M_{cv}}{M_{wp} + M_{cv}}$$

where
  F_{sf} = the RAM sampling factor
  M_{cv} = the mass of water discharged to the conditioning vessel
  M_{wp} = the mass of water discharged to waste

Passing the sieve

Calculate the mass of particles passing the 150-micron sieve using the fines correction value. Add this figure to the previous two gradings to obtain the complete aggregate grading.

1.7.3 Pressure filter (Sandberg) method

Recover the aggregate from the test sieves and sieve to obtain a grading. Determine the mass of particles passing the 150-micron sieve using the fines correction factor and combine with the previous figures to obtain the aggregate grading.

1.8 Summary

This chapter has detailed tests that can be carried out on fresh concrete to determine the cement, pfa, ggbs and water contents. Time is of the essence on all construction sites, so the faster test methods using RAM for testing cement content and the microwave oven method for drying samples are most advantageous from this context.

Reference

Clear, C.A. (1988) Delayed analysis of fresh concrete for cement and water content by freezing. *Magazine of Concrete Research*, **40**, No. 145, December.

2

Strength-testing machines for concrete

J. B. Newman

2.1 Introduction

For control or other purposes concrete is normally required to be tested under uniaxial compression (cubes, cores or cylinders), indirect tension (cylinders) or flexure (prismatic beams). In the UK the tests are required to conform to British Standard Specifications for uniaxial compression (BS EN 12390-3 and BS 1881: Part 116), indirect tension (BS EN 12390-6 and BS 1881: Part 117) and flexure (BS EN 12390-5 and BS 1881: Part 118). Direct tension testing is not standardized but a number of testing machines have been produced for research purposes.

For indirect tension and flexure the testing machine specification is given in the appropriate standard which also refers to calibration details. However, compression-testing machines have been the subject of more rigorous standardized perfomance testing in view of the importance of concrete cube, core and cylinder core and cylinder testing within the construction industry.

2.2 Uniaxial compression testing

2.2.1 Introduction

Figure 2.1 shows the results of a survey to identify the effects of deviating from the standard procedures for determining cube strength and indicates the importance of the

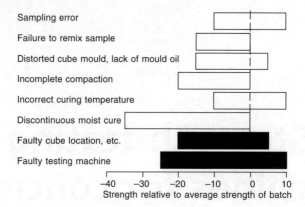

Figure 2.1 Effect of deviations from standard procedures for determining cube strength.

method of test. Prior to about 1972 machines for testing cubes in accordance with BS 1881: Part 2: 1970 were required essentially only to satisfy the load calibration requirements of BS 1610. For various reasons, mainly connected with the design of individual machines, this resulted in a wide disparity of results for the same concrete tested by different testing machines throughout the UK (Concrete Society, 1971) (Figure 2.2).

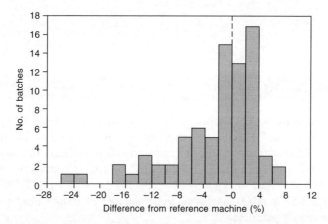

Figure 2.2 Comparative cube tests on pre-1972 Grade 2 compression testing machines.

In 1965 a 'round robin' series of tests conducted by a well-known firm of consultants showed the means of sets of 12 cubes of 'identical' concrete sent to 8 laboratories to vary by up to ±10 per cent. A later (1972) series organized by the BRMCA (Stillwell, 1972) showed that more than 50 per cent of the 105 testing authorities included in the survey (representing approximately one third of approximately 300 testing authorities in the UK) produced results significantly different from those obtained using a reference machine *with some more than 30 per cent different*! The summarized results from this survey are given in Table 2.1.

The conclusion was drawn that '... at least half the testing authorities in the country should be investigating their curing procedures, their testing procedures, their testing machines, or all three'.

Table 2.1 Comparison between the results of cubes tested by various laboratories and a reference compression testing machine

Type of lab.	No. of reports	Mean strength difference between lab and reference m/c (%)	% of results		
			<4% different	≥4% different	≥10% different
Commercial testing	31	5.8	31	61	19
Major contractor	7	4.0	71	29	0
Readymixed concrete supplier	5	3.3	80	20	0
Cement manufacturer	12	5.3	50	50	8
University	10	5.0	40	60	10
Polytechnic	6	3.1	83	17	0
College of technology	30	7.1	37	63	17
Local government	4	5.6	46	54	12

The introduction of the draft 'New Unified Code of Practice' for the structural use of concrete based on probabilistic principles (which eventually became CP 110:1972 and then BS 8110) provided another impetus for change. It was thus decided to attempt to reduce the range of results for a concrete of a given grade to acceptable limits (say, less than ±5 per cent).

In 1971 a small Working Party was set up by the Concrete Society with members from BRMCA, IC and C&CA to draw up clauses which could be used as a basis for a British Standard on machines for cube testing. Close cooperation with the testing machine manufacturers was maintained throughout the deliberations of the Working Party and in 1972 a Concrete Society Working Party report entitled 'Main clauses for inclusion in a standard for concrete cube compression testing machines' was produced (Concrete Society, 1972). This report identified a number of facets of machine design which could cause errors in the measured failure load in addition to load calibration errors. Such factors included:

- Load indication
- Lateral and axial stiffness of frame (Figure 2.3)
- Spherical seating design (Figure 2.4)
- Machine platen shape and hardness etc.
- Auxiliary platen shape and hardness
- Specimen location

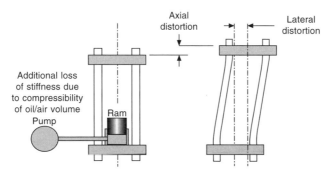

Figure 2.3 Problems due to axial and lateral stiffness of testing machine.

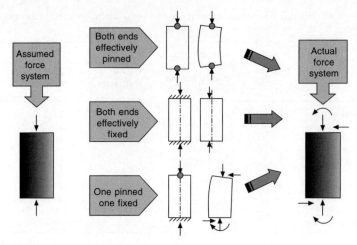

Figure 2.4 Problems with force transfer to specimens.

The report rapidly became adopted by the manufacturers as a performance 'standard' to be used for the design of new machines. Eventually the BSI adopted the recommendations given in the Working Party report in a revision of BS 1881 (BS 1881-115).

Despite the advances in testing machine design and verification it is essential to observe and record the mode of failure of concrete specimens (Figure 2.5).

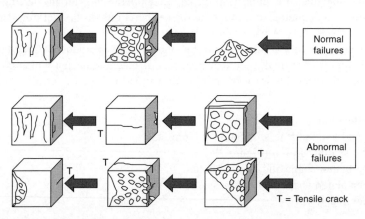

Figure 2.5 Normal and abnormal cube failures.

One-sided (bending) modes of failure are associated with reduced strengths and can be caused by one or more of the following faults:

- placing cubes off-centre and faults in cube moulds (Figure 2.6)
- faults in testing machine (loose or misaligned parts)
- insufficient lateral stiffness of the testing machine
- premature locking of the ball seating before mating with cube
- rotation of the ball seating under load
- excessive lateral movement of the hydraulic ram

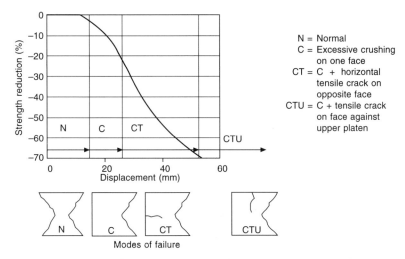

Figure 2.6 Strength and failure type for 100 mm cubes placed eccentrically in testing machine.

- excessive lack of planeness of loaded surfaces of platens or specimen (Figure 2.4) with the influence on fracture and strength depending on the quality (stiffness) of the concrete (RMC Technical Services, personal communication) (Figure 2.7)
- fins on cube (particularly at later ages)

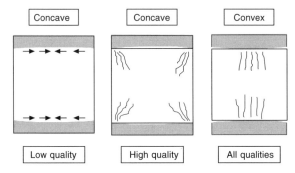

Figure 2.7 Fracture modes for cubes of different qualities of concrete tested using non-plane platens.

The effect on measured cube strength of the behaviour illustrated in Figure 2.5 is given in Table 2.2.

2.3 Specification for compression testing machines

BS EN 12390-4 covers requirements for three classes of machine and the principal items are outlined in Table 2.3. BS EN 12390-4 is essentially a performance specification which allows each manufacturer to design a testing machine using his own expertise and ideas.

Table 2.2 Effect of platen planeness defects on measured cube strength of concrete of differing quality

Platen defect	Quality (stiffness) of concrete	Effect on cube strength	Reason
Concave (Dished)	Low	Increased	Increased lateral restraint
	High	Decreased	Stress concentration near edges
Convex (bowed)	Any	Decreased	Stress concentration near centre and decreased lateral restraint

Table 2.3 BS EN 12390-4 requirements for compression testing machines

Force	Control	Smoothly and without shock Manual or automatic
	Pacer	Accuracy ±5%
	Indicator	Analogue or digital display with maximum load indication Calibrate after installation, annually or when relocated or repaired Accuracy: Class 1 ±1%, Class 2 ±2%, Class 3 ±3% Use only upper 80% of range for each indicator
	Application	Apply force through ball-seating and main machine platens or auxiliary platens
Platens	Machine	Upper platen shall incorporate a ball-seating Specified minimum hardness to ISO 6507-1 Specified flatness tolerance to Annex A of prEN 12390-1 Specified surface texture Area at least as great as area of specimen
	Auxiliary	As for machine plates with additional requirements: Opposite faces parallel to ±0.05 mm Thickness at least 23 mm
Spacing blocks		Up to 4 blocks may be inserted to reduce distance between machine platens Circular or square cross-section of 200 mm minimum diameter or length of side Adequately supported from *below* Locate positively and accurately on vertical machine axis Flatness and parallelism as for auxiliary platens
Specimen location		Locate accurately using positive physical location if required
Performance requirements		Centre of rotation of ball-seating shall coincide with centre of upper machine platen and be able to rotate at least 3° Component parts of m/c shall be aligned accurately Upper machine platen shall align freely with specimen or auxiliary platen on initial contact and then lock under load
Design for verification		Accuracy of force indication Self-alignment of upper machine platen Alignment of component parts Restraint of movement of platen

2.4 Verification procedures

2.4.1 Force transfer

For new machines the characteristics of force transfer are measured using a strain-gauged column ('Footemeter') or equivalent device as described in Annex A of BS EN 12390-4. The 'Footemeter' consists of a 100 mm diameter 200 mm long steel cylinder of specified requirements for type of steel and dimensional tolerances. The device is fitted with matched temperature-compensated electrical resistance strain gauges (ERSG) in four complete bridges measuring axial and circumferential strain. The strain gauges are measured using a dedicated eqipment. The device is shown in Figures 2.8 and 2.9.

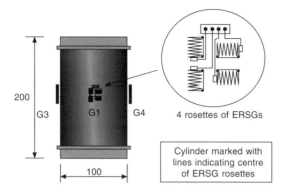

Figure 2.8 Strain-gauged column.

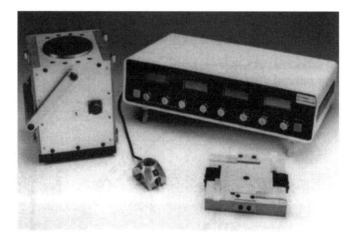

Figure 2.9 Strain-gauged column and strain measuring equipment.

The verification procedure checks the following:

- Self-alignment of the upper machine platen and machine component parts
- Self-alignment of the upper machine platen
- Alignment of component parts of machine
- Restraint on movement of upper machine platen

2.4.2 Force calibration

The force calibration equipment shall conform to EN 10002-3 and the procedures are described in Annex B of BS EN 12390-4. The flatness of the machine and auxiliary platens and the rate of force application are also checked.

2.4.3 Comparative cubes

The procedures to be adopted to provide an effective verification of a concrete cube testing machine in operation are described in BS 1881 Part 127. The procedure compares the cube test results obtained using the machine with those obtained on a reference machine using a combination of cube sizes and strength levels that will disclose deficiencies. The method for manufacturing and curing the cubes is highly controlled. Verification should be carried out at intervals required by the relevant accreditation authority (see Tables 2.4 and 2.5).

Table 2.4 Definitions

Reference laboratory	Accredited by UKAS to provide cubes and cube testing
Reference machine	Compression testing machine used only for reference purposes
Batch	Quantity of concrete mixed in one cycle of operations of mixer
Group	Total number of cubes of one size produced from same batch in moulds filled and vibrated together
Set	Six cubes of one size produced from same batch but from six different groups
Long-term coefficient of variation	Mean coefficient of strength of latest 25 consecutive sets of cubes of same nominal size and strength tested on the reference machine

Table 2.5 Procedures

Specimens	Three sets of 6: cubes as follows: 150 mm with mean strength 70–85 MPa 100 mm with mean strength 70–85 MPa 100 mm with mean strength 14–19 MPa
Cube density	Mean density for each set should differ from mean density of the batch by <5 kg/m^3 and the standard deviation (SD) of density for the batch should be <6 kg/m^3
Calculation	Mean strength and the SD for each set of six cubes on reference machine and machine being verified. Discard results *for batch* if the SD is >3.0 MPa for high-strength cubes or >0.7 MPa for low-strength cubes.
Analysis of performance	(i) Difference between mean strength of each set of cubes tested on reference machine and machine being verified as % of mean strength on reference machine (ii) 95% confidence interval for difference as % of mean strength on reference machine
Performance criteria	(a) For each of three sets of cubes difference between means <4.0% (b) Where the SD for high-strength cube set on machine being verified exceeds 3.0 MPa it should not exceed 3.3 × SD of corresponding results on reference machine (c) Where the SD for the low-strength cube set on the machine being verified exceeds 0.7 MPa it should not exceed 3.3 × SD of corresponding results on reference machine

2.5 Tensile strength testing

Direct tension testing is carried out for research purposes using various types of testing machine specially developed to apply, as far as possible, uniaxial tensile force to concrete specimens. Such machines include those applying force through bonded end plates or through 'lazy tong' grips (Figure 2.10). However, such procedures are relatively complex, induce secondary restrain stresses and require expensive equipment in addition to the compression test facilities normally available in concrete testing laboratories.

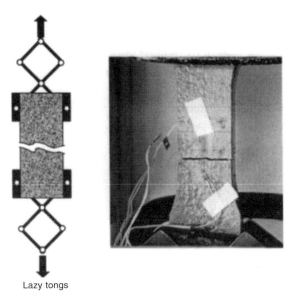

Lazy tongs

Figure 2.10 Direct tension applied by 'lazy tongs' device.

Fortunately, the particular characteristics of concrete under multiaxial states of stress results in the tensile strength as derived from the indirect tension test being close to the direct tensile strength 6, Volume 2. In the tensile splitting test (also known as the indirect tension or 'Brazilian' test), described in BS EN 12390-6 a cylinder is placed between the platens of a standard compressive testing machine with its axis horizontal. Force is applied along a vertical diameter through thin plywood or hardboard packing strips interposed between the cylinder and the machine platens. Normally, a jig ensures correct alignment and a compression testing machine to the requirements of BS EN 12390-4 is used. Failure occurs by splitting along the loaded diameter predominately under a state of biaxial compression/tension with the maximum tensile stress of $2P/\pi DL$ where P is the applied load and D and L are the diameter and length of the cylinder respectively. The maximum tensile stress at failure is the splitting or indirect tensile strength. In fact, in the zone of the loaded diameter near the centre of the cylinder, where failure is considered to initiate, the near-biaxial state of stress is as follows (Figure 2.11):

Vertical compression (σ_1)	3 units
Axial stress (σ_2)	Approximately zero
Horizontal tension (σ_3)	1 unit

That the indirect tensile strength is close to the direct tensile strength is due to the shape of the biaxial compression/tension strength envelope (Figure 2.12).

Zones 1, 2 and 3 in Figure 2.12 denote failures in compression/tension, near uniaxial compression and biaxial compression, respectively. The figure indicates that the tensile

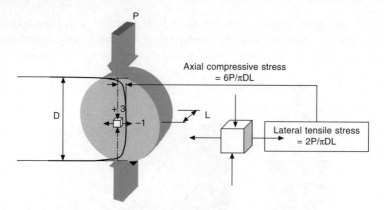

Figure 2.11 Stress state on element at centre of indirect (splitting) tensile test specimen.

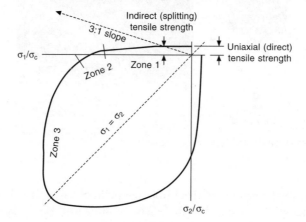

Figure 2.12 Diagram showing a typical biaxial strength envelope for concrete and the stress trajectory near the centre of the test specimen in an indirect (splitting) tensile test.

stress at failure under a compression/tension ratio of –3 is approximately the same as the tensile stress at failure under uniaxial tension.

2.6 Flexural strength testing

A prismatic beam of concrete is supported on a steel roller bearing near each end is loaded through similar steel bearings placed at the third points on the top surface (2-point loading). Test details are described in BS EN 12390-5. Normative Annex A describes a method whereby the load is applied through a single roller at centre span (centre-point loading).

For two-point loading a constant bending moment is produced in the zone between the upper roller bearings. This induces a symmetrical triangular stress distribution along vertical sections (assuming elasticity) from compression above the neutral axis at mid-height to tension below the neutral axis. The flexural strength (the maximum tensile stress at the bottom surface) is FL/bd^2 where F is the total load, L is the distance between the lower supporting rollers and b and d are the breadth and depth of the beam. The Standard gives details of the testing rig and requires that the compression testing machine used to apply load shall conform to BS EN 12390-4.

For centre-point loading the flexural strength is $3FL/2bd^2$ which has been found to give results 13 per cent higher than two-point loading.

References

BS 1610: Materials testing machines and force verification equipment.
BS 1881-115: Testing concrete: Specification for compression testing machines for concrete.
BS 1881: Part 116: Testing concrete: Method for determination of compressive strength of concrete cubes.
BS 1881: Part 117: Testing concrete: Method for determination of tensile splitting strength.
BS 1881: Part 118: Testing concrete: Method for determination of flexural strength.
BS 1881: Part 127: Testing concrete: Method of verifying the performance of a concrete cube compression machine using the comparative cube test.
BS 8110-1: Structural use of concrete: Code of practice for design and construction.
BS CP 110: The structural use of concrete (superseded by BS 8110).
BS EN 12390-3: Testing hardened concrete: Compressive strength of test specimens.
BS EN 12390-4: Testing hardened concrete: Compressive strength: Specification for testing machines.
BS EN 12390-5: Testing hardened concrete: Flexural strength of test specimens.
BS EN 12390-6: Testing hardened concrete: Tensile splitting strength of test specimens.
BS EN ISO 376: Metallic materials: Calibration of force-proving instruments used for the verification of uniaxial testing machines.
Concrete Society (1971) The performance of existing testing machines, Report PCS 62, London, July.
Concrete Society (1972) Main Clauses for Inclusion in a Standard for Concrete Cube Compression Testing Machines, 3rd report of a Concrete Society Working Party, Concrete Society.
ISO 6507-1: Metallic materials: Vickers hardness test: Test method.
prEN 12390-1: Testing hardened concrete: Shape, dimensions and other requirements of specimens and moulds.
Stillwell, J. (1972) BRMCA figures show test house failings. *New Civil Engineer*, **31**, August.

3

Accelerated strength testing

Tony Binns

3.1 Aims of accelerated and early-age testing

The principal aim of accelerated and early-age testing of concrete is to allow early predictions to be made of the 28-day strength. Changes in concrete properties may be detected at an early stage and appropriate corrective action taken to improve concrete quality. This action is intended to avoid the production and use, over periods of several days and weeks, of concrete that is either sub-standard or over-designed. It is important to avoid both extremes, where the cement content and/or strength of the concrete are significantly lower than – or significantly greater than – the optimum for performance and conformity to specifications.

In the former case, savings can be made by minimizing risks arising from contractual claims based on non-conformity of sub-standard concrete. The costs of investigation, assessment and remedial work, including consequential costs surrounding the effects on other site activities and the delay in the construction programme, can be considerable.

In the latter case, where concrete is over-designed, cost savings can be made by maintaining economical mix designs through early warning that strength is higher than expected. Downward adjustment of strength may be possible within a few hours of producing the concrete. This facility means that savings in cement content may be made earlier than would be possible with a regime based on testing after many days of normal curing at a standard temperature of 20°C, for example.

Accelerated or early-age testing can replace 28-day testing for purposes of routine quality control, specification and conformity testing.

3.2 Principles

The properties and performance of concrete, including (but not only) strength development are difficult to predict because of variability in environmental conditions and the quality of constituent materials. While constant environmental conditions and uniform materials can be found in laboratories and test-houses, concrete on site and in precast concrete factories may be subjected to a range of exposure conditions in which the temperature and humidity will affect concrete properties and construction schedules. The environment will similarly affect test specimens that are made and left adjacent to the work they represent. Materials supplied in bulk are subject to variations in quality that might not be detected until standard specimens are tested, normally after 7 and 28 days. Accelerated testing both enables reliable, early predictions to be made of concrete cube strength by eliminating environmental factors and also allows detection of materials-based influences on concrete performance.

A number of methods of accelerated and early-age testing have been developed that allow early predictions to be made of the behaviour of concrete at later ages. Accelerated construction techniques require accelerated testing regimes if the quantity of concrete at risk is to be minimized. Fast-track construction programmes and the use of high-performance concrete call for early removal of formwork on site and, in precast concrete factories, early demoulding that subject concrete to early stresses. Response to stress at early age needs to be understood if failures are to be avoided and this is made possible by monitoring and predicting long-term strength at an early stage of its development.

Testing programmes based on multiples of 7 days are most convenient if weekend work is to be minimized, because specimens are made and tested on the same day of the week. A 28-day lunar cycle for testing was chosen as a suitable regime consistent with convenience and the time scale necessary for adequate strength development due to the relatively slow rate of hydration of early concretes.

In accelerated and early-age testing a sample of concrete is divided into two or more parts: one part is subjected to an early-age test and the other(s) tested at a later age to establish a relationship between early and later age test results.

The relationship that is found to exist between early and later age test results enables a correlation factor to be calculated. This correlation factor is used for the prediction of later age test results based on those obtained from accelerated or early-age testing.

Correlation is reaffirmed or adjusted on the basis of continuous monitoring of the relationship between early age and later age test results by ensuring that every early-age specimen is matched by an identical specimen made at the same time under the same conditions from the same sample of concrete and cured under standard conditions until the time it is tested, normally 28 days.

Accelerated and early-age testing are most often applied to the control of 28-day strength (Figure 3.1). Provided corresponding relationships can be established between properties of concrete at an early age at other critical times, the methods can – or could – also be used for predicting:

Formwork striking times
Handling of pre-cast concrete units
Elastic modulus, expansion, shrinkage and creep
Abrasion resistance
Longer-term strength development

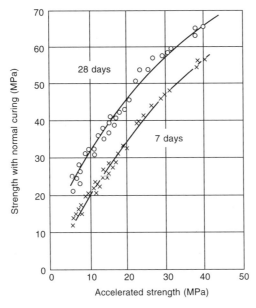

Figure 3.1 Typical relationship between accelerated test results and standard-cured 7- and 28-day strengths.

3.3 British Standards procedures

3.3.1 General procedures

BS 1881: Part 112 (BSI, 1983a) provides for three alternative methods based on different curing regimes, giving strength results on the day following casting. All three methods have the following procedures in common:

- Test cubes are cast as normal in accordance with BS 1881: Part 108 (BSI, 1983b)/BS EN 12390–1 (BSI, 2000a).
- A cover plate is fitted to the top of the mould to isolate the concrete from the water in the curing tank.
- The cubes are immersed in water in an insulated curing tank fitted with a lid.
- The cubes are demoulded and tested in accordance with BS 1881: Part 116 (BSI, 1983c)/BS EN 12390–3 (BSI, 2000b).

Timing is critical for the procedures.

35°C method
- Immediately after casting and sealing, cubes are immersed in water at 35°C ± 2°C for 24 hours ± 15 minutes.
- Demoulding and referencing of the cubes is completed in not more than 15 minutes.
- The cubes are tested in accordance with BS 1881: Part 116 (BSI, 1983c)/BS EN 12390–3 (BSI, 2000b) immediately on removal from tank.

55°C method

- After casting and sealing, cubes are left undisturbed at 20°C ± 5°C for a period of time between 1 hour and 1 hour 30 minutes.
- In not less than 1.5 hours and not more than 3.5 hours after mixing, the cubes are immersed in water at 55°C ± 2°C for not less than 19 hours 50 minutes.
- Demoulding and referencing the cubes is completed by not more than 20 hours 10 minutes after mixing.
- The cubes are immersed in water in a cooling tank at 20°C ± 5°C for 1 to 2 hours.
- The cubes are tested in accordance with BS 1881: Part 116 (BSI, 1983c)/BS EN 12390–3 (BSI, 2000b) immediately after removal from the cooling tank.

82°C method

- After casting and sealing, cubes are left undisturbed at 20°C ± 5°C for at least 1 hour (no maximum limit is given).
- The cubes are placed in an empty tank which is then filled with tap water at 5°C to 20°C (see Figure 3.2).
- The water temperature is raised to 82°C ± 2°C within 2 hours ± 15 minutes and maintained at this temperature for 14 hours ± 15 minutes.
- The hot water is drained off within 15 minutes.
- The cubes are removed, demoulded and referenced immediately.
- The cubes are tested in accordance with BS 1881: Part 116 (BSI, 1983c)/BSI EN 12390–3 (BSI, 2000b) immediately while still hot.

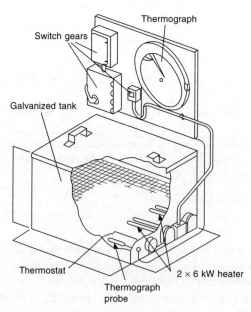

Figure 3.2 Typical curing tank, 82°C method, BS 1881: Part 112.

3.3.2 Test report – mandatory information

(a) Identification of the specimen
(b) Method and temperature of curing
(c) Maximum and minimum recorded curing temperatures
(d) A certificate confirming that the test has been performed in accordance with the Standard.

3.3.3 Test report – optional information

(a) Time of adding water to the mix
(b) Time of making the cubes
(c) Time of immersion in the curing tank
(d) Time of removal from the tank
(e) Time cubes placed in, and removed from, the cooling tank (82°C method only)
(f) Temperature record during curing.

3.4 American Society for Testing and Materials (ASTM) procedures

3.4.1 General procedures

ASTM C684 (ANSI, 1999) provides for four alternative methods based on different curing regimes, giving results after 5 hours, 24 hours, 28.5 hours or 49.5 hours after casting.

Procedure A – Moderate heating method (24 h at 35°C)
- Moulds are warmed up before calibration, casting and testing.
- Specimens (cylinders) are cast and sealed by fixing a plate (such as a machined baseplate) to the top of the mould.
- Sealed specimens in their moulds are immersed in water at 35°C ± 3°C for a total of 23.5 h ± 30 min.
- The specimens are tested for strength in a compression testing machine after 24 h ± 30 min from the time of casting.

Procedure B – Thermal acceleration method (boiling water)
- Moulds are warmed up before calibration, casting and testing.
- Specimens (cylinders) are cast and sealed by fixing a plate (such as a machined baseplate) to the top of the mould.
- The sealed specimens are immersed in their moulds in water at 21°C ± 6°C for 23 h ± 15 min.
- The sealed specimens are immersed in their moulds in boiling water for 3.5 h ± 5 min.
- The sealed specimens are carefully removed from the water and allowed to cool in air for not less than 1 hour.

- The specimens are tested for strength in a compression testing machine after 28.5 h ± 15 min from the time of casting.

Procedure C – Autogenous curing method

The word autogenous relates to a self-generating process (such as in autogenous healing or autogenous shrinkage). In the context of accelerated testing the heat of hydration liberated during the exothermic reaction between cement and water is exploited as a method for raising the temperature of concrete specimens without the need of any external source of heat.

- Specimens (cylinders) are cast and sealed by fixing a machined plate or tight-fitting cap to the top of the mould.
- The sealed specimens in their moulds are placed inside a plastic bag and stored for not less than 12 hours in air at 21°C ± 6°C.
- The temperature of the specimens is checked then they are placed, in their moulds, inside an insulated container made from rigid thermal insulation material whose insulation meets precise heat retention requirements specified in the Standard (Figure 3.3). The insulated container closely surrounds the moulded specimens, includes a maximum/minimum thermometer and has an outer casing and an inner liner to protect it from mechanical damage.

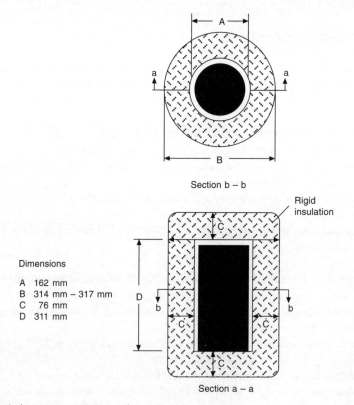

Dimensions
A 162 mm
B 314 mm – 317 mm
C 76 mm
D 311 mm

Figure 3.3 Typical apparatus, ASTM procedure C, Autogenous curing.

- The specimens are removed from the container after 48 h ± 15 min from the time of casting.
- The specimens are demoulded and tested for strength in a compression testing machine at 49.5 h ± 15 min from the time of casting.

Procedure D – Elevated temperature and pressure (K-5) method
- Specimens (cylinders) are cast in special stainless steel cylinder moulds each equipped with a thermometer and a heating element capable of raising the concrete temperature within the mould to 150 ± 3°C. The moulds are equipped with removable top and bottom metal plugs fitted with O-ring seals.
- Top surface of the concrete is screeded to accept the metal plug that transmits pressure.
- Three specimens are stacked vertically on top of each other in a special loading rig inside an insulated container.
- A stress of 10.3 N/mm^2 ± 0.2 N/mm^2 is applied by hydraulic hand pump and maintained
- The temperature is raised by the heating elements to 150°C ± 3°C within 30 ± 5 min.
- The elevated temperature is maintained for 3 h ± 5 min and pressure maintained for a total of 5 h ± 5 min
- The specimens are extruded from their moulds and tested for strength in a compression testing machine within 15 min of demoulding. (See Figures 3.4 and 3.5.)

3.5 Other national standards

Standard procedures for accelerated testing currently exist in the following countries:

- Australia (Standards Australia 2000)
  AS 1012.19.1-2000. Methods of testing concrete – Accelerated curing of concrete compression test specimens – hot water method.
  AS 1012.19.21-2000. Methods of testing concrete – Accelerated curing of concrete compression test specimens – warm water method.
- Canada (Canadian Standards Agency)
  CSA A23.1. Accelerating the cure of concrete cylinders and determining their compressive strength.
- Denmark (Dansk Standard)
  DS 423.26. testing of concrete – Hardened concrete – Compressive strength – Accelerated curing for 24 hours.
- Russia (State Committee of the Russian Federation of Standardization and Metrology)
  GOST 22783-77. Concretes – Method of accelerated determination of compressive strength.

3.6 Other research

Work published in 2000 in Thailand (Thanokom *et al.*, 2002) found that accelerated test results can be produced within 3.5 to 5.5 hours (depending on the fineness of cement) with the use of microwave energy. The paper includes a state-of-the-art review by Kasai of a range of methods developed and reported in Japan for concrete strength prediction. The review outlines the following methods:

3/8 Accelerated strength testing

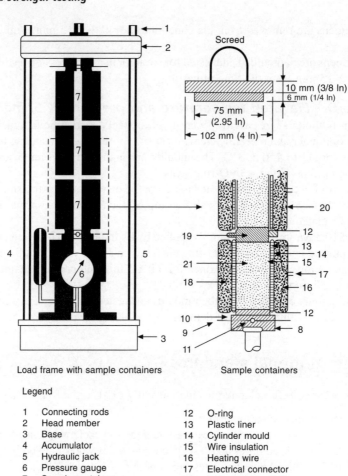

Figure 3.4 Typical K-5 test apparatus, ASTM C 684 Procedure D.

- Relationship between strength development and the unit amounts of cement, water and free water/cement ratio.
- Accelerated strength tests on concrete and mortar by
  - Hot water curing
  - High temperature curing in sealed restricted moulds
  - Set-accelerating admixtures.

3.7 Applications of accelerated and early-age testing

It is recognized by concrete producers that a delay of 28 days from mixing concrete to testing is an unacceptably long time. However, although accelerated testing, using hot

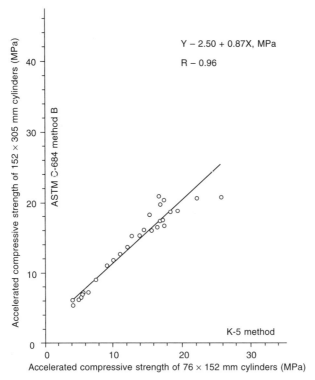

Figure 3.5 Relationship between ASTM Procedures B (boiling water method) and D (K-5 method).

water, was popular with UK ready-mixed concrete producers in the 1970s, it currently appears to be little used.

Accelerated and early-age testing are suitable for all projects in which concrete is used. They are appropriate to all types of concrete production and construction where a very early indication of strength is required.

Materials engineers on large contracts monitor and take action on early-age test results that can include results from high-temperature accelerated procedures. Precast concrete manufacturers use early-age cube results for safe de-moulding, pre-stressing, lifting, stacking and transporting of structural precast elements, often based on the strength of cubes cured alongside the elements they represent. The precise procedures for curing specimens under the same conditions as the elements they represent are not specified in any national Standard but are written in manufacturers' method statements and quality manuals.

Many civil engineering contractors use 'early-strike' test cubes which means they are cured adjacent to – and under the same general conditions as – the work they represent in order to save valuable time on the construction programme. Because test specimens are compacted differently from *in situ* concrete and any benefit derived from heat of hydration in large elements is unlikely to be replicated inside a small steel cube mould, uncertainties surrounding the relevance of 'early-strike' test results exist, due to these differences in compaction techniques and curing temperatures.

The majority of ready-mixed concrete producers in the UK currently use 7-day early-age prediction of 28-day compressive strength test results in their routine procedures of quality control and conformity testing, using standard curing tanks filled with clean, circulated water at a temperature of 20°C ± 2°C. The predictions are compared with target and/or specified strengths and early corrective action is taken to control mean strength and cement content for a given concrete strength.

If concrete were to be tested at an age of less than 7 days, the British Standard (BSI, 1983d) requires the tolerance on water temperature to be ± 1°C. Twenty-eight-day strengths are predicted by empirical correlations based on known historical data for each type of concrete or by established maturity functions.

Accelerated testing techniques can also be used to enable other properties of mature concrete to be predicted and adjusted as necessary, thereby making it possible to avoid deterioration or excessive movements in hardened elements. Accelerated test rigs have been successfully used for predicting shrinkage, creep and elasticity and accelerated tests include methods for predicting the resistance of concrete to freeze–thaw attack, acid attack, corrosion and abrasion.

3.7.1 Control by prediction of 28-day strength

As early as 1915 research by McDaniel was leading towards prediction of 28-day strength but no feasible procedure was developed. A large amount of data was produced in 1933 during the construction of the Hoover Dam in USA and Patch (1933) proposed an eight-hour accelerated test method as a routine quality control technique. Malhotra (1981), however, reported that Patch's method had been found unreliable by the US Bureau of Reclamation. In 1958 King commended accelerated testing to the 50th annual conference of the Institution of Civil Engineers in London. The principles of King's procedure were:

- Specimens to be cast in moulds sealed with greased plates to prevent drying and placed in an airtight oven within 30 min of casting.
- The temperature in the oven to be raised to 93°C within 1 hour and maintained for 5 to 6 hours after casting.
- Cubes to be demoulded, referenced and allowed to cool in air at room temperature for 30 min.
- Cubes to be tested for strength in a compression testing machine at 7 hours after casting.

Because of inconvenience of timing, modifications were made to permit a longer pre-heating time (normally 18 to 24 hours over 16°C water) followed by a shorter 4th hour period in the oven, giving results the day following casting.

In 1963 in Ontario, Smith and Chojnacki developed the 'fixed set' boiling water method:

- Specimens (cylinders) are cast in standard cylinder moulds as normal.
- A condition referred to as 'Fixed set' is determined by Proctor needle penetration resistance (quoted as 24 MPa) normally 6 to 8 hours after mixing.
- The moulds are sealed by fixing a plate (such as a machined baseplate) to the top of the mould.
- The specimens in their moulds are immersed in boiling water within 20 min of determining the fixed set condition.

- The specimens remain immersed in boiling water for 16 hours.
- The specimens are demoulded, referenced and capped with sulphur or calcium aluminate cement mortar within 60 min.
- Crushing test as normal.

The fixed set method was included with autogenous curing and boiling water methods in Canadian national standard CSA A23.1 in 1970 but has not been continued in current revisions of the Standard (Canadian Standards Agency).

In 1966 Grant (Grant and Warren, 1977) presented a paper to a London Symposium on Concrete Quality, introducing the method based on heating concrete cubes to 82°C that became incorporated into BS 1881: Part 112 (BSI, 1983a) and gave rise to the adoption of accelerated testing by the UK ready-mixed concrete industry for many years.

Some major producers decided it was advantageous to establish a secure relationship between accelerated and 28-day test results in order to predict the likely 28-day strength in less than 24 hours and, in UK, the 55°C and the 82°C methods of BS 1881: Part 112 were in widespread use from about 1970. At this time, the influence of different cement types and admixtures became better understood (Figure 3.6).

Besides being wasteful of large quantities of very hot water, the 82°C method also represented a health and safety risk and was abandoned by gradual phasing-out in the last decade of the twentieth century.

The experience of accelerated testing gained through routine use of the methods described in BS 1881: Part 112 by many UK readymixed concrete producers enables results to be entered directly into statistical control programmes such as Cusum for control of cement content, standard deviation and the 7-day: 28-day correlation. It will be seen elsewhere in this book that standard deviation is controlled by monitoring the range between each pair of predicted 28-day strengths. If the average range of pairs of consecutive predicted 28-day strengths is susbstantially different from the target range, then the standard deviation must have changed and corrective action may be called for, usually in a recalculation of the target mean strength which in turn represents a shift in cement content.

Correlation factors are established for a range of different concretes from 7- and 28-day test data. The relationship over a range of concretes is non-linear, represented by a polynomial curve; that is, the best-fit curve approximating to a number of fixed points, usually of the order of three. This gives an enhanced prediction of low-strength concretes but a less reliable prediction for high-strength concretes.

It is essential that the assumed 7:28 day strength relationship is continuously checked for accuracy. An initial correlation factor is established as the target value then, as more 7- and 28-day test results become available, their relationship is compared against the target. Any substantial variation from the target correlation indicates that the assumed value is inaccurate and needs to be adjusted. The effects of changing the prediction of 28-day strengths also brings changes to the control of mean strength and standard deviation relative to their target values.

Statistical analysis by Cusum techniques permits relatively easy and sensitive control of these interdependent target values of predicted strength.

The behaviour of concrete in response to the application of heat is dependent upon the mineralogical composition of the cement type including any additions, the chemistry of any admixture and the strength development characteristics of the concrete at the standard temperature (e.g. 20°C). Consequently the correlation factor is unique to a single reference concrete unless a modification is possible, which would link the reference concrete to

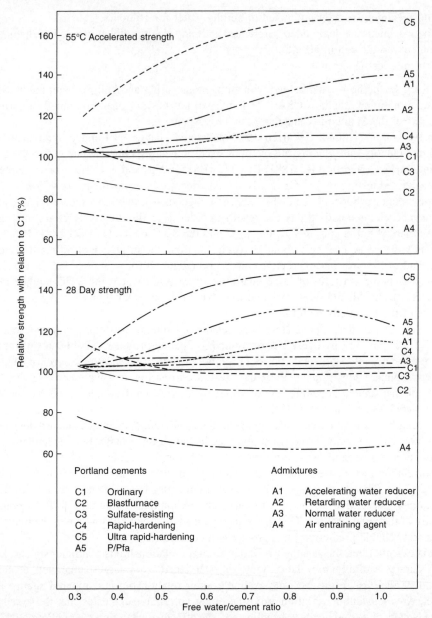

Figure 3.6 Effects of different cement types and admixtures on correlation of accelerated and 28-day strengths.

others (typically concretes made from the same materials but with different cement contents) by applying a further correlation established by comparative testing, this time between the reference concrete and another.

Specific correlation factors for predicting 28-day strength from 7-day test results apply to each source and type of cement and for a finite period only. This makes the technique most suitable for large construction projects, manufacture of precast concrete in factories and the production of ready-mixed concrete where materials sources are likely to be constant. The correlation factor is unaffected by all but extreme changes in aggregate type.

As more data become available general strength relationships can be derived and the formulae in Table 3.1 were proposed in 1983 by Dhir and Gilhespie (1983).

Table 3.1

Material groupings	28-day strength (f_{c28}) – N/mm² f_{ca} is accelerated strength, 55°C method
Plain concrete	$f_{c28} = 2.411 - 0.0202\,(f_{ca})^2$
Concrete with water-reducing admixture	$f_{c28} = 2.359 - 0.0191\,(f_{ca})^2$
Air-entrained concrete	$f_{c28} = 2.024 - 0.0120\,(f_{ca})^2$
PC + Pfa	$f_{c28} = 2.512 - 0.0227\,(f_{ca})^2$

Little work has been published on effects of ggbs on the correlation of accelerated or early-age strength with standard 28-day strength, although some Chinese work (Zhengyong and Shiqi) on mortar sieved from concrete containing cement made with 40% ggbs was reported in 1989. In this procedure the mortar was dosed with a set-accelerating admixture and strength predictions were possible after only one hour.

The relationship between 28-day and accelerated strength flattens at high concrete strength levels due to the concrete reaching a strength limited by the aggregate strength. In most concretes, aggregate particles are stronger than the cement matrix surrounding them. This is confirmed during cube testing when it is rare to see much damage to aggregates. However, in the quest for higher compressive strength, a reduction in free water/cement ratio produces a matrix with a strength that will eventually exceed that of the aggregate being used. Aggregate particle shape and texture also influence the relationship between early strength, 28-day strength and ceiling strength where the bond between aggregates and mortar matrix if smooth rounded gravels is lower than that of granular or crystalline crushed rocks. It is standard industry practice to make an automatic application of an appropriate 'aggregate strength factor' to allow for such effects (Figure 3.8).

3.7.2 Conformity

Where a proven relationship exists between accelerated/early-age strength and standard 28-day strength, formal contractual acceptance of conformity could be based on the early test results. The actual 28-day test results would serve to confirm the quality of concrete represented by the test data and, thereby, the validity of the early results. Any action deemed necessary in the event of non-conformity, including investigative procedures, can commence without delay (Figure 3.9).

CP 110 (BSI) recognized and permitted the use of accelerated and early-age testing in specification and conformity criteria. Canada, Denmark, the UK, the USA and the former USSR are among those nations who have, at some stage, adopted accelerated testing in their standards which could be used for quality control purposes. However, no references

Accelerated strength testing

Cement source A		Cement source B		Cement source C	
7-day	28-day	7-day	28-day	7-day	28-day
5	7.0	5	9.5	5	6.5
6	8.5	6	12	6	8.0
7	10.0	7	14	7	9.0
8	11.5	8	16.5	8	10.5
9	13.0	9	18.5	9	11.5
10	14.5	10	20.5	10	13.0
11	16.0	11	22.5	11	14.5
12	17.5	12	24.5	12	15.5
13	18.5	13	26.5	13	17.0
14	20.0	14	28.5	14	18.0
15	21.5	15	30.0	15	19.5
16	22.5	16	32.0	16	20.5
17	24.0	17	33.5	17	22.0
18	25.5	18	35.5	18	23.0
19	26.5	19	37.0	19	24.0
20	28.0	20	38.5	20	25.5
21	29.0	21	40.5	21	26.5
22	30.5	22	42.0	22	27.5
23	31.5	23	43.5	23	29.0
24	33.0	24	45.0	24	30.0
25	34.0	25	46.5	25	31.0
26	35.5	26	48.0	26	32.5
27	36.5	27	49.5	27	33.5
28	37.5	28	50.5	28	34.5
29	39.0	29	52.0	29	35.5
30	40.0	30	53.5	30	37.0
31	41.5	31	54.5	31	38.0
32	42.5	32	56.0	32	39.0
33	43.5	33	57.0	33	40.0
34	44.5	34	58.5	34	41.0
35	46.0	35	59.5	35	42.5
36	47.0	36	61.0	36	43.5
37	48.0	37	62.0	37	44.5
38	49.0	38	63.0	38	45.5
39	50.0	39	64.5	39	46.5
40	51.0	40	65.5	40	47.5
41	52.5	41	66.5	41	48.5
42	53.5	42	67.5	42	49.5
43	54.5	43	69.0	43	51.0
44	55.5	44	70.0	44	52.0
45	56.5	45	71.0	45	53.0
46	57.5	46	72.0	46	54.0
47	58.5	47	73.0	47	55.0
48	59.5	48	74.0	48	56.0
49	60.5	49	75.0	49	57.0
50	61.5	50	76.0	50	58.0
51	62.5	51	76.5	51	59.0
52	63.5	52	77.5	52	60.0
53	64.5	53	78.5	53	61.0
54	65.5	54	79.5	54	62.0
55	66.5	55	80.0	55	63.0
56	67.5	56	81.0	56	64.0
57	68.5	57	82.0	57	65.0
58	69.5	58	83.0	58	66.0
59	70.5	59	84.0	59	66.5
60	71.0	60	84.5	60	67.5
61	72.0	61	85.5	61	68.5
62	73.0			62	69.5
63	74.0			63	70.5
64	75.0			64	71.5
65	76.0			65	72.5
66	77.0			66	73.5
67	77.5			67	74.5
68	78.5			68	75.5
69	79.5			69	76.0
70	80.5			70	77.0
Aggregate strength factor	120	Aggregate strength factor	120	Aggregate strength factor	120

Figure 3.7 Working Cusum correlation tables. [Note: Aggregate strength factor shown indicates the use of crushed rock coarse aggregate; the equivalent value for gravel aggregate by this method is currently 80.]

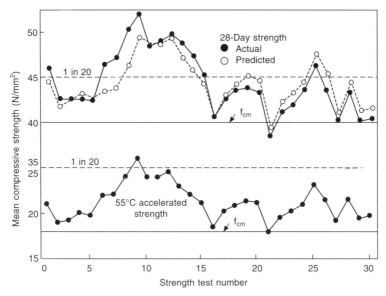

Figure 3.8 Accelerated versus standard-cured 28-day strengths: groups of four consecutive test results.

exist in BS 8110 (the successor to CP 110), BS 5328, BS EN 206-1 or BS 8500. BS EN 12390 does not currently include any procedure for accelerated testing techniques.

Specialized standards, including BS 5502, BS 7777, BS 8004, BS 8007 cross-refer to BS 8110 and/or BS 5328 for details of conformity testing and, accordingly, do not make reference to accelerated or early-age techniques.

Where accelerated testing has been used in contract specifications and conformity clauses, caution has normally dictated that normally-cured specimens should also be tested at a later age.

Projects on which accelerated or early-age testing has been either specified or permitted include:

M6 Midland Link Motorway, Gravelly Hill Interchange, Birmingham, UK
CN Communications Tower, Toronto
Deep drainage system, Mexico City
La Angostura hydroelectric project, Mexico
Al Shwarif regulating tanks, Great Man-Made River Project, Libya
Channel Tunnel Rail Link, UK

3.8 Conclusion

There is a strong case to be made for abolishing the specification of concrete strength on the basis of tests carried out at 28 days in favour of high-temperature accelerated testing or tests at 7 days. Where a proven relationship exists between accelerated/early-age strength and 28-day strength, formal acceptance of conformity may be made as soon as the early test results are made available. The actual 28-day test results would serve to confirm the conforming status of the early results. In the event of poor correlation, factors that may interfere with strength characteristics – for example, lack of uniformity in aggregates or curing conditions – may be identified and corrected.

Because of intrinsic differences in early-age performance of concretes, design strength will probably still need to be specified in terms of 28-day strength class or, possibly, strength at later ages such as 56 and 90 days but, in an age of fast-track construction and high-performance concrete, it appears wasteful to disregard the depth of knowledge and experience that exists regarding early-age testing.

It is worthwhile considering whether concrete specifications can readily and safely allow conformity testing to be based on early-age strength results linked to an agreed correlation continuously monitored by actual 28-day strength test results. Accelerated testing, based on the use of elevated temperatures, suffers the disadvantages of requiring (a) strict discipline for control of timings and temperatures that does not lend itself readily to typical construction or production schedules and (b) occasional weekend testing. 7-day testing at standard temperature is convenient and, subject to the establishment of appropriate correlation, can be used for routine production control and for conformity testing.

References

ANSI (1999) ASTM C684-99 Standard test method for making, accelerated curing and testing concrete compression test specimens. American National Standards Institute.
BSI (1983a) BS 1881: Part 112: 1983. Testing concrete – Methods of accelerated curing of test cubes. British Standards Institution, London.
BSI (1983b) BS 1881: Part 108: 1983. Testing concrete – Method for making test cubes from fresh concrete.
BSI (1983c) BS 1881: Part 116: 1983. Testing concrete – Method for determination of compressive strength of concrete cubes.
BSI (1983d) BS 1881: Part 111: 1983. Testing concrete – Method of normal curing of test specimens (20°C method).
BSI (2000a) BS EN 12390–1: 2000. Testing hardened concrete: Shape, dimensions and other requirements for test specimens and moulds.
BSI (2000b) BS EN 12390–3 Testing hardened concrete: Compressive strength of test specimens.
BSI BS 5328 – Concrete: Part 4 Specification for the procedures to be used in sampling, testing and assessing compliance of concrete. British Standards Institution, London.
BSI BS 5502 Buildings and structures for agriculture. British Standards Institution, London.
BSI BS 7777 Flat-bottomed, vertical, cylindrical storage tanks for low temperature service. British Standards Institution, London.
BSI BS 8004 Code of practice for foundations. British Standards Institution, London.
BSI BS 8007 Code of practice for design of concrete structures for retaining aqueous liquids. British Standards Institution, London.
BSI BS 8110. The structural use of concrete. British Standards Institution, London.
BSI BS 8500 Concrete – Complementary British Standard to BS EN 206-1. British Standards Institution, London.
BSI BS EN 206 – Concrete: Part 1: Specification, performance, production and conformity. British Standards Institution, London.
BSI CP 110. Code of practice for the structural use of concrete. British Standards Institution, London.
Canadian Standards Agency. CSA A23.1 Accelerating the curing of cylinders and determining their compressive strength.

Dansk Standard, The Danish Standards Association. DS 423.26 Testing concrete – Hardened concrete – Compressive strength – Accelerated curing for 24 hours.

Dhir, R.K. and Gilhespie, R.Y. (1983) Concrete quality assessment rapidly and confidently by accelerated strength testing. European Ready Mixed Concrete Organization, London, May.

Grant, N.T. and Warren, P.A. (1977) Forecasting concrete strength. *Civil Engineering*, July–August.

King, J.W.H. (1958) Accelerated testing of concrete. 50th Annual Conference, Institution of Civil Engineers, London, May.

Malhotra, V.M. (1981) *Concrete International*.

McDaniel, A.B. (1915) Influence of temperature on the strength of concrete. Bulletin No. 81, July, University of Illinois Engineering Experiment Station.

Patch, D.G. (1933) An 8-hour accelerated strength test for field concrete control. *Journal ACI Proc.* 4–5 March–April.

Thanakom, P., Cayliani, L., Dumangas, M.I. Jr and Nimityongskul, P. (2002) Prediction of later-age compressive strength of normal concrete based on accelerated strength of concrete cured with microwave energy. *Cement & Concrete Research*, 32.

Smith, P. and Chojnacki, B. (1963) Accelerated strength testing of concrete cylinders. *Proc. ASTM*, 63.

Standards Australia. AS 1012.19.1-2000. Methods of testing concrete – Accelerated curing of concrete compression test specimens – hot water method. AS 1012.19.2-2000. Methods of testing concrete – Accelerated curing of concrete compression test specimens – warm water method.

State Committee of the Russian Federation of Standardization and Metrology. GOST 22783. concretes. Method of accelerated determination of compressive strength.

Zhengyong, C. and Shiqi, L. (1989) Accelerated strength test results in one hour with slag cement concrete in Beijing. Third CANMET/ACI international conference on fly ash, silica fume, slag and natural pozzolans in concrete. Trondheim, Norway.

4

Analysis of hardened concrete and mortar

John Lay

4.1 Aims and objectives

The purpose of this chapter is to outline the reasons for the analysis of hardened concrete and mortar. It describes the information that can be obtained from analysis and outlines the procedures for sampling and analysis. Factors influencing accuracy and precision are discussed and the interpretation of results is considered.

4.2 Brief history

Procedures for the analysis of hardened concrete have existed for many years. In 1950, National Building Studies Technical Paper No. 8 (Bowden and Green, 1950) described classical chemical methods. These methods were probably already long-established by 1950. Developments in analytical techniques led to the publication in 1971 of *The Analysis of Concretes* by the Building Research Station (Figg and Bowden, 1971). In the same year, The British Standards Institution (BSI) produced its first British Standard Methods of Testing Concrete, BS 1881: Part 6, 'in substitution for the many methods previously in use'.

In 1980, the BSI revised BS 4551, Methods of testing mortars, screeds and plasters, first published in 1970 (BSI, 1980). The procedures for chemical analysis were superior to those described in the concrete standard and were adopted for use with concrete in BS

1881: Part 124, Methods for analysis of hardened concrete in 1988 (BSI, 1988). Both standards are still in use, although the mortar methods underwent minor revision in 1998. Most British Standards are soon to be replaced by European Standards. There is currently no work programme underway to replace BS 1881: Part 124 or BS 4551: Part 2 (BSI, 1988, 1998), so they will continue to be used for the foreseeable future.

4.3 Introduction

Chemical analysis can be used to provide information on the composition of hardened concrete and mortar, and to give clues to the reasons for any deterioration. Straightforward British Standard procedures are available. Reasonable analytical chemistry skills, modest laboratory equipment and an understanding of concrete technology are required. An accurate knowledge of the chemical composition of the constituents of the concrete or mortar is desirable, but more often than not unavailable. This means that assumptions must be made based on the analyst's experience and judgement. Interpretation based upon sound assumptions is often the key skill in translating analytical results into meaningful information with a sensible level of accuracy and precision.

4.4 Reasons for analysis

The most common reason for analysis is the failure of a concrete test cube to comply with a compressive strength requirement. The failure may be due to the concrete mix proportions being outside the specified limits, or because the cube was badly made or poorly cured. Analysis to determine the cement content of the cube can establish the likely cause of failure.

If there is some doubt about the quality of concrete already placed in a structure, cores can be drilled and analysed. Concrete or mortar that has deteriorated or failed in service can be analysed to determine possible reasons for the deterioration.

4.5 Information that can be obtained by analysis

It is usually necessary to determine Portland cement content. Even if the cement content of the concrete or mortar is not itself a concern, other determinations such as chloride or sulfate content can only be put into context if the cement content is established. If a blended cement (e.g. Portland blastfurnace cement in concrete, or masonry cement in mortar) has been used, the cementitious material content can usually be determined provided the relative proportions of the blend or its actual chemical analysis are known. The relative proportions of a blend of ground granulated blastfurnace slag (ggbs) and Portland cement can usually be estimated by chemical analysis, but pulverized-fuel ash (pfa), microsilica and metakaolin can be impossible to quantify.

Original water/cement ratio of concrete can be determined if the samples are undamaged and have not undergone earlier tests involving, for example, compression. Representative samples of the aggregates used in the manufacture of the concrete are needed for the determination of free water/cement ratio. If aggregate samples are unavailable, only total water/cement ratio can be found.

Aggregate type and grading can be assessed. Grading can only be carried out when the aggregate is both strong and essentially insoluble in acid. Because of the nature of the procedures used to recover the aggregates from hardened concrete, results are indicative only and should not be used to assess compliance with particle size distribution specifications.

Chloride, sulfate and alkalis contents of hardened concrete and mortar can be determined when deterioration has occurred, or when an assessment of the risk of future problems is required. Portland cement is usually the source of most sulfate and alkalis, and cement content is critical to how much chloride can be tolerated. Tricalcium aluminate (C_3A), a constituent of Portland cement, binds chloride ions so that it is not available to induce reinforcement corrosion. It is therefore necessary to determine cement content in order to properly assess chloride, sulfate and alkali levels.

The lime content of hardened mortar can be determined. Once again, Portland cement content is assessed, and any calcareous material not accounted for by the cement is assumed to originate from lime. This can lead to inaccuracies if the sand used in the manufacture of the mortar contains calcareous material and a sample of the sand is not available to allow corrections to be made.

4.6 Sampling procedures

4.6.1 General

The most important aspect of chemical analysis is sampling. A perfect analysis of a poorly taken sample yields little information about the true composition of the concrete or mortar. Analysis may be required for a variety of reasons, to investigate a range of problems, and to provide information on widely varying masses of concrete or mortar. It is therefore important that the interested parties agree the method of sampling and the quantity of material represented before the sample is taken. Some general rules can be applied to the mass of sample required (BSI, 1988, 1998):

1. The minimum linear dimension of the sample must be at least five times the nominal maximum dimension of the largest aggregate particles.
2. A minimum mass of 1 kg of concrete is required. For the determination of original water/cement ratio, at least 2 kg are required, and 4 kg are needed if aggregate grading is to be assessed.

   A minimum mass of 100 g of mortar or render is required. Sub-samples should comprise at least 50 g. Sub-samples should not represent more than 10 m^2 of masonry, and the entire sample should not represent more than 50 m^2 of masonry. Each sub-sample of screed should weigh at least 100 g, and the entire sample should not represent more than 100 m^2.
3. All foreign matter and reinforcement must be avoided unless they form part of the test. If original water/cement ratio is to be determined, the sample should be in one piece, with no cracks or damage. This means that cubes or cores that have been tested in compression are not suitable.

The sampling strategy should also be considered. It may be that the average quality of a large quantity of concrete is under investigation. Or perhaps the quality of a particular part of the element is in question, and a comparison of satisfactory and suspect concrete is required. Sampling locations will be different in the two cases, because unrepresentative or damaged areas should be avoided if average quality is of interest. In any case, all interested parties should agree sampling locations.

The concrete in the top 10 per cent and bottom 10 per cent of a lift may be unrepresentative. Research has shown that in walls and columns, cement content at the top can be much higher than average, and cement content at the bottom can be lower than average. Highly workable concrete is more likely to show this variation (Skinner, 1980). In walls and columns the top and bottom of the lift should be avoided for representative sampling. It is still necessary to include them when sampling slabs and beams, however.

4.6.2 Sample types

If the sample presented for analysis is a test cube, then the whole cube constitutes the concrete sample. This means that if the cube has been tested in compression, all fragments should have been recovered and included with the sample, and any fragments from other cubes must be excluded. It is essential that the fresh concrete used to make the cube was sampled in accordance with British Standard procedures (BSI, 1983).

The best method of sampling from a hardened concrete element is by cutting cores (Concrete Society, 1989). It is usually necessary to take 100 mm diameter cores for 20 mm aggregate concrete and 150 mm diameter cores for 40 mm aggregate. Vertically drilled cores should be the full depth of the concrete, and horizontally drilled cores should be the full width. There may of course be situations where this is not possible.

When coring is impractical, it may be possible to break out a lump of concrete. The sample should be a single piece of sufficiently large size. If the lump breaks up on removal from the concrete element, all fragments must be collected to form the sample. Full-depth samples should be obtained from slabs. Debris or lumps that have apparently fallen from the structure should not be used. The location of the sample (and therefore the concrete that it represents) can only be established by actually breaking it out of the element. Lumps are not suitable for the determination of original water/cement ratio.

For the measurement of chloride, dust drillings are suitable. Several holes should be drilled at each location using a slow-speed hammer drill with a 10 mm to 25 mm bit (Lees, 1995). At least 30 g of material should be collected from each location.

The preferred method of sampling mortar is to remove a whole masonry unit and take the full depth and thickness of the bedding mortar as the sub-sample. It is important to avoid contamination with fragments from the masonry units. Raking mortar from the bed is usually unacceptable, since it only yields the weathered surface mortar unless it is done to considerable depth. Mortar from the perpend (vertical) joint should not be taken because it may not be representative (Lees, 1995).

Render (mortar used to coat a wall) should be sampled by cutting out an area of at least 150 mm by 150 mm to form the sub-sample. Screed (mortar used to coat and level a floor) should be sampled by cutting out or coring full-depth pieces over the floor area. Adhering substrate must be removed from render and screed samples before analysis.

Although mortar, render and screed sub-samples are usually combined for analysis, the sub-samples should be clearly identified and separated for examination before analysis.

It is, for example, possible that in an area of masonry, there may be several batches of mortar of different composition. In this case, separate analysis of the sub-samples is required. Render samples may consist of a number of layers. It may be necessary for the analyst to painstakingly separate the layers for individual analysis. BS 4551 gives details of mortar, screed and render sampling (BSI, 1998).

4.6.3 Number of samples

Concrete is a variable material, even when it is correctly batched, mixed, placed and compacted. Sampling and testing inevitably introduce more variability. As the number of samples taken increases, the sampling error is reduced. If the samples are then analysed separately, testing error is reduced and some assessment of the variability of the concrete in the element is obtained. If the samples are combined before analysis, sampling error is reduced but testing error is unchanged and no information on variability of the concrete is gained.

When sampling from a small volume of concrete, say a single truckload of 6 m^3, at least four independent samples should be taken (Concrete Society, 1989). Table 4.1 shows the errors associated with taking fewer samples.

Table 4.1 Errors arising from concrete variability in an element

Number of independent samples	Range of error in determined cement content arising during sampling (kg/m^3)
1	±50
2	±35
3	±30
4	±25

Note that the errors in the table are calculated at a 95 per cent level of confidence, assuming a standard deviation of 25 kg/m^3 for the cement content of the concrete. The values do not include any allowance for errors arising during analysis.

When sampling from a larger volume of concrete or from a large number of similar units, ten to twenty independent samples should be taken at random (Concrete Society, 1989). The samples should be examined and analysed individually to assess the extent of any problem. Areas identified as requiring further investigation can then be treated as small volumes of concrete.

A sample of mortar, screed or render should consist of at least five increments. These sub-samples can be combined for analysis (if careful examination shows no appreciable differences between them). Each sub-sample should represent no more than 10 m^2 of masonry or floor area. The combined sample should not represent more than 50 m^2 of masonry for mortar and render, and should not represent more than 100 m^2 of floor for screed (BSI, 1998).

4.6.4 Sample preparation

After visual examination in the laboratory, samples are reduced to fragments and dried. The sample is divided in a representative way, for example by riffling or quartering. Approximately a quarter of the sample is retained for future tests or reference. The remaining three-quarters is successively crushed and reduced to produce a sub-sample passing a 1.18-mm sieve. This sub-sample is in turn ground and reduced to provide an analytical sample passing a 150-μm sieve.

Approximately 5 g of the analytical sample will be used in each replicate determination of cement content. It is vitally important that these 5 g remain representative of the sample as a whole. This means that no fine material should be lost during crushing and grinding, and nothing should be allowed to contaminate the sample. All sample reduction should be carried out with the utmost care, because it is so substantial – a 2.4 kg concrete test cube (itself representative of several tonnes of concrete) is reduced almost five hundredfold to provide the test portion. Figure 4.1 illustrates the task.

4.7 Determination of cement content of concrete

The principle behind the determination of cement content is straightforward. The analyst should select a chemical compound that constitutes a large proportion of the cement and is essentially absent from the other components of the concrete. The concrete is then analysed for that component. If all the component originated from cement, and the proportion of the component in cement is known, then cement content can be calculated.

For example, Portland cement typically contains 64.5 per cent calcium oxide. If a concrete contains 13 per cent Portland cement and 84 per cent aggregate containing no calcium oxide (the other 3 per cent is combined water of hydration), then the concrete will contain:

$$\text{Calcium oxide content of concrete (\%)} = 13 \times 64.5/100 \tag{4.1}$$

Therefore if the calcium oxide content of the concrete has been determined, the cement content can be calculated as:

$$\text{cement content of concrete (\%)} = \text{calcium oxide content of concrete} \times 100/64.5 \tag{4.2}$$

The cement content in kg/m^3 can be calculated by multiplying the percentage cement content by the dry bulk density of the concrete.

In reality, most aggregates contain a proportion of the chemical compound that is analysed to determine cement content. This has two main implications. First, the equation to determine cement content must be adjusted to allow for the component in the aggregate:

$$\text{Cement content (\%)} = 100 \times (c - b)/(a - 1.23 \times b) \tag{4.3}$$

where
a = percentage of the component in the cement
b = percentage of the component in the aggregate
c = percentage of the component in the concrete sample

A 14 tonne truckload of concrete is sampled to make cubes.

Each cube weighs about 2.4 kg. The cube is successively crushed and reduced to produce the analytical sample.

Each portion used for chemical analysis weighs around 5g.

Figure 4.1 The scale of sample reduction for chemical analysis (after Lees, 1995).

Second, because the actual analysis of the aggregate used in the manufacture of the concrete is rarely known, the cement content should be determined a second time using another major constituent of the cement. In practice, soluble silica is the only truly suitable alternative to calcium oxide. Soluble silica typically makes up around 20 per cent of Portland cement.

The concrete is analysed to determine soluble silica and calcium oxide contents, and the cement content is calculated from the two determinations. If the two results are within 1 per cent of each other, the mean value is reported. If not, reasons for the discrepancy are investigated. If a reason is found then the preferred value is reported. For example, if the aggregate contains a large proportion of calcareous material, the result based on calcium oxide content may be unreliable, so the result based on soluble silica content is preferred. If no reason is found then the lower value is preferred.

The extraction method used for the British Standard (BSI, 1988) determination of cement content involves an acid dissolution step and an alkaline dissolution step. The acid extraction dissolves all the calcium oxide from the concrete sample, and some of the soluble silica. The alkaline extraction dissolves the rest of the soluble silica originating from the cement. Siliceous aggregates such as flint consist largely of insoluble silica. The extraction procedures are designed to dissolve the cement completely with minimum dissolution of the aggregate. However, excessively fine grinding of the analytical sample, increased alkaline extraction time or excessive alkaline extraction temperature can all lead to some dissolution of the aggregate silica. This may lead to errors in the determined cement content.

Any calcareous material in the sample originating from the aggregate will be dissolved by the acid extraction step. It is desirable to analyse samples of the aggregate used in the manufacture of the concrete in order to correct for this source of calcium oxide. However, problems can still arise if the aggregates contain small but variable amounts of calcareous material.

Portland cement from a known source is generally uniform in its chemical composition. It is usually sufficient to use the manufacturer's monthly average composition for the month of manufacture, but control samples can be analysed to provide data for the calculation of cement content. The manufacturer's composition may also be used for ground granulated blastfurnace slag.

However, the declared composition of pulverized-fuel ash is not suitable for use in calculations. The silica within pulverized-fuel ash converts from an insoluble to a soluble form during the course of the pozzolanic reaction. The extent of this conversion is variable and unpredictable. It is therefore best to calculate the cement content of a concrete containing pulverized-fuel ash from the calcium oxide content, since the conditions of extraction can be expected to dissolve all the calcium oxide regardless of the extent of the pozzolanic reaction (Figures 4.2–4.4).

4.8 Analysis of mortar to determine mix proportions

The principles behind the analysis of cement:sand mortar, render and screed are essentially the same as those for concrete. When a cement:lime:sand mortar is analysed, the cement content is calculated from the soluble silica determination and any calcium oxide unaccounted for by the cement is used to calculate the hydrated lime content. If the sand contains any

Analysis of hardened concrete and mortar

Figure 4.2 Separation of silica from the concrete extracts.

Figure 4.3 Crucibles containing silica are ignited to constant mass at 1100°C.

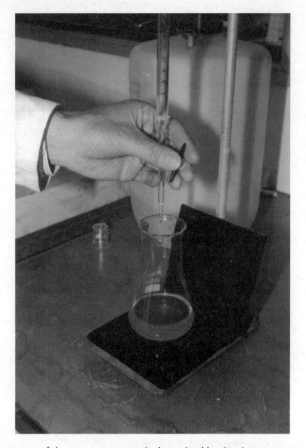

Figure 4.4 Calcium content of the concrete extracts is determined by titration.

calcareous material, reliable sand samples are needed to allow the calculation of mix proportions.

Because most mortars are specified by volume, it is necessary to convert the mass proportions determined from the analysis to volume proportions. This can be a source of errors because the dry bulk densities of the constituents of mortar may vary. However, BS 4551: Part 2 contains standard assumed bulk densities for use when accurate values are not available. BS 4551: Part 2 allows a range of cement and lime contents for each mortar mix designation, so in practice the effect of the bulk density of constituents on the calculations is minimized.

4.9 Other determinations

4.9.1 Determination of sulfate content

All the sulfate in concrete or mortar is extracted during the acid extraction stage of the determination of cement content. Consequently, sulfate content of a portion of the extract can be determined. If cement content is not being determined, a portion of the powdered concrete or mortar can be extracted into acid and the sulfate content determined.

4.9.2 Determination of chloride content

Portions of powdered concrete are extracted in boiling nitric acid and the chloride content is determined titrimetrically. Sub-samples of the extracts from cement content determinations are not suitable for the determination of chloride, because hydrochloric acid (itself a source of chloride ion) is used in the cement content extraction.

4.9.3 Determination of alkalis content

Portions of powdered concrete are extracted in boiling nitric acid and the sodium oxide and potassium oxide contents are determined by flame photometry. The procedure can give unduly high results, however. Many aggregates contain sodium and potassium compounds that are not available in concrete. However, the fine grinding of the concrete sample can allow these alkalis to be extracted. It may be possible to make an allowance for such occluded alkalis when assessing the results.

4.9.4 Determination of original water/cement ratio of concrete

Concrete contains water for two principal reasons, first, to hydrate the cement and second, to make the concrete workable enough to compact. The water of hydration chemically binds in the cement hydrates. The water in the concrete that was added to provide workability occupies a system of capillary pores within the concrete. In order to determine original water/cement ratio it is necessary to determine the combined water in the cement hydrates, the capillary porosity and the cement content. The sum of the combined water and the capillary porosity is equivalent to the original water content.

Combined water content is a very difficult determination to carry out, and the result must be corrected for combined water in the aggregates. The British Standard (BSI, 1988) allows an assumed combined water content of 23 per cent of the cement content to be used if the determination is not carried out.

Capillary porosity is determined by drying a concrete sample and refilling the empty capillary pores with a dense organic liquid. In order to determine original free water/cement ratio, the capillary porosity of the aggregates must be determined and the result corrected. If reliable aggregate samples are not available, only the original total water/cement ratio can be determined.

The test is not suitable for mortars, or air-entrained, cracked, or poorly compacted concretes

4.9.5 Aggregate grading

A concrete sample is carefully broken down and the cement is dissolved in hydrochloric acid. The recovered aggregate particles are washed and graded.

The test is only really suitable for aggregates that are insoluble in acid. Calcareous coarse aggregates can be separated manually from the cement: sand matrix, but the process is laborious and may not be reliable. Weak aggregates cannot be analysed by this method.

Because of the nature of the separation procedure, a variable amount of aggregate breakdown occurs during the test. This usually leads to an overestimate of the fineness of the coarse aggregate (due to breakdown), and may lead to an underestimate of the fineness of the sand (due to the presence of coarse aggregate debris) (Concrete Society, 1989). It is not usually possible to make a reliable correction for the breakdown. This means that the gradings obtained are useful for comparative purposes or to give an indication of particle size distribution, but they cannot be used to assess compliance with BS 882 or other specifications (BSI, 1992).

Mortar samples can also be tested to determine recovered sand grading. The procedures are similar to those described for concrete, but a hot acid breakdown is used, followed by simmering in hot sodium bicarbonate solution.

4.9.6 GGBS content

Ground granulated blastfurnace slag (ggbs) is usually the only constituent of concrete that contains appreciable quantities of sulfide. The sulfide content of the concrete can be determined, and if the sulfide content of the ggbs used in the manufacture of the concrete is known, the ggbs content of the concrete can be calculated. It is necessary to have an accurate analysis of the ggbs in order to account for the soluble silica and calcium oxide present in the concrete originating from the ggbs. This allows the calculation of the Portland cement content.

Sometimes, not all the sulfide in the concrete is recovered, perhaps due to oxidation during sample preparation. This can be detected from discrepancies in the determined Portland cement contents. No published precision data is available, but experience suggests a repeatability of 15–20 per cent of the mean may be appropriate. However, experience of alternative methods of determining ggbs content such as electron microscopy suggests that they suffer from similar levels of precision.

Pulverized-fuel ash (pfa) has similar soluble silica and calcium oxide contents to flint sand when fresh, and has a variable composition in concrete depending on the extent of the pozzolanic reaction. With low calcium oxide pfa (such as those found in the United Kingdom), the ash does not significantly interfere with the determination of Portland cement content by calcium oxide.

Microsilica and metakaolin are rapidly consumed in concrete by the pozzolanic reaction. Their presence may be betrayed in cement content analysis by an unusually high soluble silica content. Again, this does not significantly interfere with the determination of Portland cement content by calcium oxide.

4.9.7 Carbon dioxide

Carbon dioxide content of concrete and aggregate can be determined by decomposing carbonates with acid and trapping the evolved carbon dioxide on a granular absorbent.

This determination may become more important as the assessment of the risk of thaumasite sulfate attack is more frequently required.

4.9.8 Admixtures

Admixtures are present at very low concentrations in hardened concrete, ranging from about 0.003 per cent for surfactants in air-entrained concrete to perhaps 0.3 per cent for stearates in 'water-proofed' concrete. There are no standard test methods available and the reliability of some proposed methods is questionable. Even with control samples of admixture and experienced analysts, blind trials carried out by RMC have suggested that results can be of very limited value.

4.10 Accuracy and precision of determined cement content of concrete

A precision experiment was carried out in the early 1980s to assess the British Standard method for the determination of cement content of concrete (BSI, 1998; Concrete Society, 1989). Eighteen laboratories took part. The repeatability, r, was found to be 40 kg/m^3. The reproducibility, R, was 60 kg/m^3. The results from two laboratories were rejected as outliers. If they had been included, the reproducibility would be much higher. This means that there is no difference, statistically, between results obtained on the same sample by the same analyst in a short period of time if the results differ by not more than 40 kg/m^3. It also means that there is no difference between results obtained on the same sample by different laboratories if the results differ by not more than 60 kg/m^3.

The accuracy of the cement content result depends largely on appropriate aggregate corrections being applied, either from analysis of representative aggregate samples or the use of valid assumptions. If the contribution from aggregates to the results of analysis is ignored, then the likely outcome is positive bias leading to an overestimate of cement content. Other errors in accuracy can arise from the analysis of the cement and the dry bulk density used to convert the cement result from a percentage to kg/m^3.

4.11 Accuracy and precision of determined mix proportions of mortar

A precision experiment was carried out in the late 1980s to assess the British Standard method for the determination of the cement and lime content of mortars and screeds (BSI, 1998). Control samples of sand were used in the experiment. The repeatability, r, for cement content based on soluble silica was estimated to be 2.8 per cent by mass for Portland cement:lime:sand mortar. This compares with a mean determined cement content of 9.6 per cent by mass. Reproducibility, R, was 3.1 per cent. The repeatability and reproducibility of the determined lime content were found to be 2.1 per cent and 3.0 per cent, at an overall average lime content of 5.5 per cent.

The repeatability and reproducibility for cement content based on soluble silica for cement: sand screed were found to be 2.6 per cent and 3.9 per cent respectively. Based on calcium oxide, the values came down to 0.8 per cent and 1.7 per cent. However, cement content can only be calculated from calcium oxide if the mortar does not contain lime or calcareous sand.

Mortars are usually specified by volume and a range of cement and lime contents are allowed for each designation. Work carried out by RMC Readymix has found similar levels of precision to those reported above. But because the precise cement and lime content is not critical to assigning mortar designation, the RMC trial found that in 35 of the 36 mortar analyses, mix designation was correctly identified (Lay, 1994).

4.12 Summary

Hardened concrete and mortar can be analysed to provide information on original mix proportions. Analysis can also indicate possible reasons for deterioration, or can be used to assess the risk of future problems. Sampling procedures, sample types, sample size, the number of samples and sample preparation are all vitally important if the results of analysis are to be meaningful. Where possible, samples of the aggregates used in the manufacture of the concrete or mortar should be analysed alongside the concrete or mortar to allow corrections to be made for soluble constituents in the aggregates.

Precision experiments have demonstrated an acceptable level of precision for the determination of cement content in concrete and mix proportions of mortars. Results should be interpreted in the light of the available precision data.

Acknowledgements

It would not be possible to complete a chapter on chemical analysis in this book without acknowledging the work of the late Tim Lees. This chapter is largely based upon his lectures.

References

Bowden, S.R. and Green, E.H. (1950) National Building Studies Technical Paper (8). HMSO, London.
British Standards Institution (1980) BS 4551 – Methods of testing mortars, screeds and plasters. BSI, London (superseded).
British Standards Institution (1983) BS 1881 – Testing concrete. Part 101 – Methods of sampling fresh concrete on site. BSI, London.
British Standards Institution (1988) BS 1881 – Testing concrete. Part 124 – Methods for analysis of hardened concrete. BSI, London.
British Standards Institution (1992) BS 882 – Specification for aggregates from natural sources for concrete. BSI, London.
British Standards Institution (1998) BS 4551 – Methods of testing mortars, screeds and plasters. Part 2 – Chemical analysis and aggregate grading. BSI, London.
Concrete Society Technical Report No. 32. (1989) *Analysis of Hardened Concrete – A guide to tests, procedures and interpretation of results*. The Concrete Society, London.

Figg, J.W. and Bowden, S.R. (1971) *The Analysis of Concretes.* HMSO, London.
Lay, J. (1994) *An Assessment of the Precision of Analysis of Hardened Mortar.* RMC Readymix, Thorpe Confidential Technical Report No. 260.
Lees, T.P. (1995) Unpublished Advanced Concrete Technology Lecture.
Skinner, M.G. (1980) *Chemical Analysis of Hardened Concrete – an investigation of within batch variation and its effect on unit cement content.* Institute of Concrete Technology, Crowthorne, Advanced Concrete Technology Project.

Further reading

There are very few books dealing with the chemical analysis of hardened concrete and mortar. Concrete Society Technical Report No. 32, *Analysis of Hardened Concrete* is one of the few. It provides an excellent guide to test procedures and interpretation of results. BS 1881: Part 124 and BS 4551: Part 2, although written as standards, do contain many clauses of interest to the concrete technologist as well as the analyst.

5

Core sampling and testing

Graham True

5.1 Introduction

This chapter sets out to describe the currently used procedures for sampling and testing concrete cores and details the necessary planning and preliminary work, locating and sampling cores, examination, measurement, preparation prior to testing, interpretation and reporting the results.

The first half of the chapter is devoted to current practice that emanates from the relevant current standards that have been in play for some time as well as recent updates. Data from research carried out in 1997 suggests that the strength gain in concrete structures is not as currently documented. These data also include cementitious blends thus increasing the database beyond the present limit of just considering Portland cement concretes.

The last part of the chapter considers the new data, new ways of assessing core properties and their influence on strength and how this data might be interpreted.

5.2 The current situation regarding standards and guidance

For the last twenty years core sampling and testing in the UK has been carried out via reference to three documents, i.e. *Concrete Society Technical Report No. 11* (1976) and the *Addendum* (1987) which is referred to as CSTR No. 11: British Standard BS 1881:Part

120:1989, and BS 6084:1981 Guide to assessment of concrete strength in existing structures. In the USA the standard used has been ASTM C.42 Standard Method of obtaining and testing cores and sawed beams of concrete.

Two new documents have recently been published, BS EN 12504-1:2000 Testing concrete in structures Part 1: Cored specimens – Taking, examining and testing in compression (this replaces part of BS 1881-120:1989) and prEN 13791 Assessment of concrete compressive strength in structures or in structural elements. (These updates have notable exclusions by omitting calculations to convert core strength to estimated *in-situ* strength as well as correction for the presence of reinforcement in core test samples.) Concrete Society Technical Report No. 11, apart from the Addendum produced in 1987, is now notably outdated. Work started in 1990 by way of planning an update which culminated in an extensive research project reported by True (1999). However, the long-awaited authoritative project report is still to be published by the Concrete Society some six years after the data had been made available. Concern has been expressed because the conclusions are significantly different from CSTR No. 11 and are more akin to data published by Watkins (1996) from work carried out in Hong Kong as well as two project reports from the South African Roads Board Research and Development Advisory Committee (1990 and 1992).

Although amended standards are available, any description of current practice must refer in the main to the current outdated version of Concrete Society Technical Report No. 11 because that report contains procedures needed to interpret core strength results. However, CSTR No. 11 must be read in conjunction with data and conclusions given in the concrete society/DETR Project Report.

5.3 Current core sampling, planning and interpretation procedures

5.3.1 Reasons for taking and testing cores

The reasons for drilling cores for strength testing are commonly to assess one or a combination of the following:

- the quality of the concrete *provided* to a construction (Potential Strength)
- the quality of the concrete *in* the construction (*In-situ* Strength) previously known as Actual Strength.
- as a check on the capacity of the structure to carry the imposed loads. This may be as a check on:
  - the actual loading
  - the design loading
  - a projected loading from a new use
- to examine deterioration in a structure due to:
  - overloading
  - fatigue (bridge structures, machine base etc.)
  - chemical reaction (ASR or chemical spillage etc.)
  - fire or explosion
  - weathering

Some of these purposes may be better suited to other forms of assessment but core testing is usually the most economical and practical.

Concrete cores, although a primary means for assessing concrete strength in a structure, are unfortunately relegated to a secondary position for assessing the strength of the Standard Moulded Cube Strength. If credible Standard Cube Test results were available and their validity not doubted, the core test would not be required, but all too often some aspect of the cube test procedure sheds doubt on the results. To be credible, cube testing requires:

- that the concrete is sampled correctly (BS EN 12350-1:2000)
- the cubes are manufactured correctly (BS EN 12390-2:2000)
- curing is carried out correctly (BS EN 12390-2:2000)
- testing is performed correctly (BS EN 12390-3:2002)

5.3.2 Planning and preliminary work before drilling cores

Any decision to take cores should be agreed by those having professional or commercial interest and should be planned during a meeting of interested parties, on-site, and include a representative of the core drilling contractor and test house. At such a meeting the following points would be covered:

- the necessity for the test and its aims. Are the cube results sound?
- evidence of the location of the suspect concrete (from site records or NDT results).
- proposed drilling locations, number and size of cores.
- ancillary work, e.g. density tests and curing history.
- strength levels required by the specification (Potential Strength) or design (*In-situ* Strength) and the action to be taken if the estimates obtained from cores are clearly greater, less or inconclusive.
- responsibility of individuals regarding execution of the work.

But note that:

- **Potential Strength** relates to the quality of the concrete used and is an estimate of the 28-day Standard Cube Strength.
- an estimate of **Potential Strength** obtained from cores is of very limited accuracy relative to the Standard Cube Strength of the batch of concrete sampled.
- *In-situ* **strength** is the strength of the concrete as it exists in the element at the time of sampling and is the end result of the quality of the supply, workmanship of the contractor and other factors up to the time of sampling.
  Remedial work has to be agreed in writing before the coring takes place and is best carried out using a proprietary repair material. The following must be considered:
- a proprietary repair material which will have a known range of properties in both the fresh and hardened state is recommended. The repair may be in the soffit of a structure where a thixotropic material will be needed in order to prevent the repair slumping or falling out before it has fully stiffened.
- sometimes it can be beneficial to pre-cast a cylinder of concrete for insertion into the core hole and bedding in cement grout or epoxy resin.
- when appearance matters, for example in an historic or readily visible location, removing

the top outer portion of the core and bedding this into the face of the core hole filled with repair material, can provide an acceptable aesthetic repair.

5.3.3 Size, number of cores, location and drilling procedures

A core diameter and length of 100 mm is generally used so as to relate directly with 100 mm test cubes. The larger the core diameter, the better. Concrete Society Technical Report No.11 recommends 150 mm diameter cores to be used whenever possible as variability due to drilling is reduced and more reliable results are obtained. It is to be preferred that the core is free of reinforcement which may limit the core diameter. BS EN 12504: 2000 infers a preferred diameter of 100 mm which is also likely to be the optimum size when 25 mm aggregate or smaller are used. It is generally accepted that the core diameter/maximum aggregate ratio should be above a minimum of 3 in order to reduce test variability to an acceptable level.

Where reinforcement congestion is found, cores with a diameter of 75 mm can be successfully used if the maximum aggregate size is 20 mm or smaller. Bungey (1979) has published data on sampling, testing and interpreting cores down to 50 mm diameter. In general, the smaller the core diameter, the less accurate the measured strength and the lower the strength obtained. More cores may be needed in order to obtain a useable result and Neville (1995) has reported that for a given precision of the estimate of strength, the required number of 50 mm cores is probably three times larger than the number judged adequate for 100 mm cores. Bowman (1980) has reported a coefficient of variation of 28.9 per cent for 50 mm cores from *in-situ* concrete compared with a value of 19.5 per cent for corresponding 150 mm cores taken from the same concrete. Small cores may be appropriate if the item or test location is confined in section and it is considered prudent to remove as little of the section as possible. It must also be borne in mind that normal compression test machines may need adaptation to test cores smaller than 90 mm diameter.

The choice of document used for guidance will influence the number of cores removed and tested. Accuracy in estimating the strength at a particular location is increased if more cores are taken. The strength estimate from a single core can be considered to lie (with 95 per cent confidence) within ±12 per cent of the strength of the concrete at that location. For n cores, the mean core strength can be considered to lie between $\pm 12/\sqrt{n}$ per cent. This is given in CSTR No. 11 for *In-situ* Strength (or Actual Strength estimates) from core test results.

Recommendation in prEN 13791 Assessment of concrete compressive strength in structures or in structural elements, suggests that an assessment should be based on specimens from at least nine cores (100 mm diameter) from one region and an assessment from a particular structural element may be based on specimens from three cores. This is increased by a factor of three if the core size is 50 mm diameter.

CSTR No. 11 covers *In-situ* Strength and Potential Strength estimates. The accuracy of *In-situ* Strength according to CSTR No. 11 has been noted above. However, the recommendation on sample numbers and accuracy regarding Potential Strength estimates is based on a minimum of four cores taken from each batch of suspect concrete and the mean estimate of Potential Strength obtained. Data from four or more cores cannot be regarded as better than ±15 per cent of the true result. Therefore, when sampling and testing for Potential Strength, in accordance with CSTR No. 11, there is little point in taking more than four cores per suspect batch because this estimate is not improved by including more core samples.

5.3.4 Location and drilling of cores

Suitable sampling points will depend on the purpose of the testing. If the purpose is to assess the condition and effect of overloading, fatigue, chemical reaction, fire or explosion or weathering on the *in-situ* strength, it is appropriate to define the area affected by an NDT method such as Schmidt Hammer or UPV testing and then to sample from that area as well as from a comparable location that has not been affected. The results will provide both a measure of the *In-situ* Strength and an indication of the reduction in strength caused by the distress. Similarly, if the purpose of the testing is to investigate suspect new concrete that has failed the standard cube testing and is built-in to a construction, the first action must be to locate the concrete by NDT means so that samples may be taken from that suspect batch of concrete to classify the material.

The height of a concrete pour has been shown, in some instances, to influence both compaction and density. Concrete strength at the base of a wall or column may be up to 25 per cent greater than at the top due to segregation and for this reason samples should not be taken from the top 20 per cent (50 mm minimum, 300 mm maximum) unless that location contains the suspect concrete. Similarly, concrete from the outer 50 mm of cover is likely to have cured under different conditions from the central bulk of the concrete and this again should not be included in the test portion of a core.

The orientation i.e. whether a core is cut horizontally or vertically into the as-placed concrete, may provide a sample strength variation. CSTR No. 11 suggests that vertically drilled cores are stronger on average by 8 per cent than horizontally drilled ones. When deciding the location, it is prudent to scan with a covermeter in order to chose a sampling point that avoids reinforcement. Final location and size of cores may well depend on rebar size and spacing.

Core drilling is best carried out by an experienced operator using water-cooled, diamond-impregnated coring barrels driven by electric or hydraulic purpose-made equipment. The rig is usually bolted firmly to the concrete and care taken over the coring rate to ensure a parallel core sample is cut perpendicular to the surface. The sample is severed by inserting and tapping a cold chisel down the annulus cut by the barrel to cause a break-off at the bottom of the drilled length. Once the samples have been cut they should be marked with a reference number, orientation mark and location from within the structure and then wrapped in polythene. Guidance on the number and location of cores to be taken is summarized in Table 5.1.

5.3.5 Visual examination and measurements

The new standard EN 12504-1:2000 which has now replaced BS 1881: Part 120: 1983 has little requirement prior to testing other than a visual examination to identify abnormalities and measurements to determine the core diameter, length, and location and diameter of any reinforcement. BS 1881: Part 120: 1983 included the previous requirements but also detailed procedures for:

Table 5.1 Guidance on the number and location of core samples

In-situ Strength assessment	Potential Strength assessment

• one per location
• accuracy
  ±12%/√n
• *n* (the number of cores) increases to reduce the region of not proven
	• four per location
	• accuracy
	  ± 15%
	• a region of 'not proven' of ± 15% must be anticipated when comparing with a spec. or required value.

Location *In-situ* strength Any location within suspect concrete. Serviceability can best be judged from the area under greatest stress.	**Location** Potential strength Each core should be representative of an equal amount of the suspect concrete.
Omit top 50 mm or 1/5 of the height whichever is greater.	Omit the top 50 mm or 1/5 of the height. Avoid reinforcement.

- visual checks on the compaction, presence of voids, honeycombing, for cracks and segregation, coarse aggregate type and particle shape.
- comparing the core surfaces with five illustrations of excess voidage, i.e. full-size photographs that provide an estimate of the excess voidage for each core.
- mass and density measurement by water displacement following a two-day soaking in water.

CSTR No. 11 contains the above as well as the following:

- examination for lack of homogeneity of concrete within and between cores.
- whether aggregates seem continuous or gap graded and any distinctive features of the fine aggregate.
- position of any cracks, drilling damage or defects and steel inclusions.
- sketch records made of the location of defects relative to the overall geometry of the core.

5.3.6 Core preparation, conditioning and testing for density, excess voidage and compressive strength

The preferred means of end preparation of test cores is high-speed wet grinding using diamond-faced grinding wheels in a rigid assembly such as a milling or grinding machine. Guidance on end preparation is given in Annex A of BS EN 12390-3:1999. Other end preparations include capping with high-alumina cement or sulfur–sand mixture. Caps should be kept as thin as possible. Weak materials such as plaster or polyester resin in the form of car body filler are not appropriate materials for end caps because they will produce a lower core strength.

In the UK it is standard practice to test cores in a saturated condition. Saturation is assumed to have occurred after 40 hours storage under water at $20 \pm 2°C$. However, if the concrete under investigation is likely to remain dry in service, it could be more appropriate to test dry cores. Density and excess voidage are detailed in CSTR No. 11 as follows:

- prior to any capping or after grinding the core ends, the core should be soaked in water

for long enough (half an hour is usually sufficient) to allow the core volume (V_u) to be determined by water displacement.
- if the core is to be capped, this should then be carried out after allowing it to dry sufficiently. Trapped air between core and capping is to be avoided. If the density of the capping material is not known then a sample is to be stored in water along with the core and its density (D_c) determined prior to core testing.
- after water storage for 40 hours, and immediately prior to compression testing, the ground or capped core should be weighed in air in a saturated surface dry condition and in water to determine its gross weight (W_t) and volume (V_t).
- the water-soaked density of the concrete in the uncapped or ground state may then be calculated from:

$$D_a = [W_t - D_c(V_t - V_u)]/V_u$$

If the core contains reinforcement, this should be retained after compressive strength testing and its weight (W_s) and volume (V_s) determined by displacement in water.

$$D_a = [(W_t - D_c(V_t - V_u)) - W_s]/(V_u - V_s)$$

Excess voidage cannot be determined with precision, but two independent methods are detailed in CSTR No. 11.

1 By comparative visual means using the photographs in CSTR No. 11
2 From density results using saturated core densities (D_a) and an agreed value for the fully compacted density (D_p), possibly from standard cube test results as follows:

$$\text{Excess voidage} = 100 \times (D_p - D_a)/(D_p - k \times 1000)\%$$

where k = a constant in the form of a fraction of the voids that are filled with water and is assumed to be 0.5

Excess voidage is the estimated voidage level above a datum voidage of 0.5 per cent that is considered typical in fully compacted concrete. It is recommended that both methods be used and an agreed overall result established by considering the relevance or each test method.

Prior to compression testing, the mean diameter is measured at the quarter points along the core axis using calipers, taking two measurements at right angles and averaging the results to obtain an overall estimate to the nearest millimetre. The length is also determined to the nearest millimetre.

Compression testing can be carried out using a compression test machine conforming to EN 12390-4:1999, in accordance with EN 12390-3:1999. The compressive strength is obtained by dividing the maximum load by the average cross-sectional area and expressing the result to the nearest 0.5 N/mm².

5.3.7 Other tests

Non-destructive testing is not included in this section. However, ultrasonic pulse velocity (UPV), dynamic modulus and Schmidt Hammer testing are tests that need to be considered when compression testing is either inappropriate or needs supplementation. Use of NDT testing following calibration against test cores tested in compression allows the whole structure in question to be scanned in order to find out how much of it is in question.

Core sampling and testing

A useful physical test that can be carried out as an alternative to compression testing is the Point Load Strength Test. This test is used to classify rock strength and has been shown by Robins (1980) to be applicable to concrete cores. The test is carried out by applying load, via 5 mm diameter point platens, across a central diameter plane. The maximum load at failure is expressed as a Point Load Strength Index and is indirectly related to the compressive strength. Naturally, the aggregate size relative to the core diameter are factors to be considered but for a particular aggregate size and core diameter, the Load Strength Index varies with cube strength for strengths greater than 20 N/mm^2.

The advantages of this test are that it is carried out using a portable, hand-operated machine and is applicable to small core samples and short core lengths. The data can also be used as a means of estimating *in-situ* tensile strength.

5.3.8 Converting core strength results to *in-situ* cube strength and potential strength

Before planning a core sampling programme, it will have been decided by all interested parties that removing and testing cores is likely to assist in gaining more information about the concrete in question. Planning will then be directed towards establishing either an estimate of the strength of the concrete in the structures, (*In-situ* Strength) or an estimate of the strength or grade of the concrete provided for the manufacture of the element (Potential Strength).

The following information is needed in order to carry out a full assessment and enable either of the two strength estimates to be derived in accordance with CSTR No. 11:

- visual assessment of the cores to determine that each is suitable for testing, representative of the bulk of the concrete from where they were sampled and that they will provide valid strength tests. It is prudent to photograph each core if there are any apparent anomalies.
- density testing by water immersion after soaking (or after drying to a suitable moisture content). (See above.)
- estimating the excess voidage both by visual comparison and from core density and potential density. (See above.) Once the excess voidage has been determined, the relevant strength multiplying factor can be determined from Table 5.2.

Table 5.2 Strength multiplying factors for specific excess voidage

Excess voidage	Strength multiplying factor
0.0	1.00
0.5	1.04
1.0	1.08
1.5	1.13
2.0	1.18
2.5	1.23
3.0	1.28
3.5	1.33
4.0	1.39
4.5	1.45
5.0	1.51

(From Concrete Society Technical Report No. 11, p. 25)

- measurement of the length and diameter of each core. See section 5.3.6.
- noting the core axis orientation relative to the way the concrete was placed, i.e. either cut vertically or horizontally from the concrete.
- noting the diameter and location of any rebar present in each core. Rebar inclusions that are parallel to the axis of a core sample can be accommodated using the following two formulae:

For a single bar

$$\text{Corrected strength} = \text{Core strength} \times [1.0 + 1.5(\phi_r/\phi_c) \times (h/l)]$$

where ϕ_r = bar diameter
ϕ_c = core diameter
h = distance of bar axis from nearer end of core
l = core length

For multiple bars

$$\text{Corrected strength} = \text{core strength} \times [1.0 + 1.5\Sigma(\phi_r \times h)/(\phi_c \times l)]$$

If the spacing of two bars is less than the diameter of the largest bar, only the bar with the higher value of $(\phi_r \times h)$ should be considered.

- assessing the curing history and the effect on strength gain. (See Figure 5.1 taken from CSTR No. 11.)
- determining the core strength from the maximum load at failure when tested at a compression test rate between 0.2 and 0.4 N/mm^2/per second.

The following two columns give the progressive steps that convert a core strength value to either an estimate of *In-situ* Strength or Potential Strength.

Estimation of *In-situ* **Strength**	**Estimation of Potential Strength**
Either	Either
Horizontal direction	**Horizontal direction**
Core strength × 2.5/(1.5 + 1/λ)	Core strength × 3.25/(1.5 + 1/λ)
Or	Or
Vertical direction	**Vertical direction**
Core strength × 2.3/(1.5 + 1/λ)	Core strength × 3.0/(1.5 + 1/λ)
Note that λ = core (length/diameter)	Note that λ = core (length/diameter)
Then applying	Then applying
• correct for steel inclusion using formula in CSTR No. 11	• correction for steel inclusion using formula in CSTR No. 11
	• correct for compaction or voidage above 0.5%
	• correction for curing history. Normal curing requires no correction.

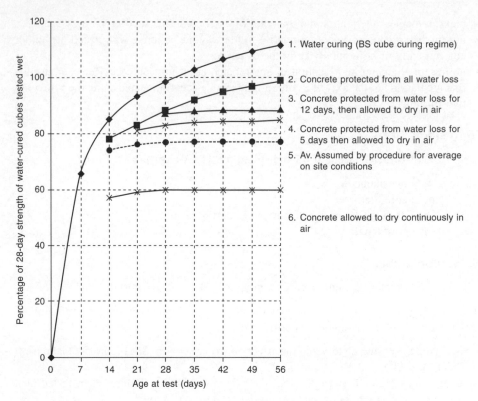

Figure 5.1 Effect of curing upon strength (determined after soaking for 2 days in water).

Hence
In-situ Strength
is obtained from:
Core strength × steel correction factor

Hence
Potential Strength
is obtained from:
Core strength × steel correction factor
× voidage correction × curing correction

5.3.9 Interpretation of results and worked examples

Concrete is considered as a randomly variable material and any collection of test results will follow a normal distribution. Differences between the results of standard cube tests and *in-situ* cube estimates will provide different distributions. Design accommodates these differences by applying a partial safety factor for strength γ_m usually taken as 1.5. Therefore it may be appropriate, depending on the design requirement, to accept concrete with a minimum *In-situ* Cube Strength of (specified characteristic strength grade)/1.5.

Potential Strength estimates carried out as described in this chapter will have a tolerance of ±15 per cent. In addition, the concrete specification may allow a minimum strength of 0.85 times the characteristic strength (C). Potential Strength estimates (P) therefore can be considered and compared as follows:

(a) If P is greater than C, there is little doubt that the concrete complies with the specification.

(b) If P is less than 0.72C (i.e. C × 0.85 × (minimum tolerance of 1 × 0.85%) = 0.85), there is little doubt that the concrete did not comply with the specification.
(c) If P is greater than 0.72C but less than C, the case is 'not proven' and engineering judgement or further tests must be applied.

prEN 13791 Assessment of concrete compressive strength in structures or in structural elements which will replace BS 1881-201 and BS 6089 provides a table that classifies *In-situ* Strength over strength classes C8/10 to C100/115 by stating that the ratio of *In-situ* to Potential Strength is 0.85, based on good curing and full compaction. No guidance is given for variations in curing and lack of compaction.

Assessment of characteristic *In-situ* Compressive Strength conformity from core tests is related to two criteria. Criterion A are applied for a series of three cores taken from one or several structural elements. Criterion B relate to one or several structural elements when at least nine cores are taken.

Criterion A
- mean of three non-overlapping test results $\geq f_{cu} + 4$
- each individual test result $\geq f_{cu} - 4$
  f_{cu} is the characteristic *In-situ* Compressive Strength taken from Table 5.3.

Table 5.3 *In-situ* Compressive Strength requirements for the strength classes according to prEN 206

Strength Class	Potential strength (N/mm²)		Ratio of *In-situ* to potential Strength*	*In-situ* Characteristic strength (N/mm²)	
	f_{cu} cylinder	f_{cu} cube		f_{cu} cylinder	f_{cu} cube
C8/10	8	10	0.85	7	9
C12/15	12	15	0.85	10	13
C16/20	16	20	0.85	14	17
C20/25	20	25	0.85	17	21
C25/30	25	30	0.85	21	26
C30/37	30	37	0.85	26	31
C35/45	35	45	0.85	30	38
C40/50	40	50	0.85	34	43
C45/55	45	55	0.85	38	47
C50/60	50	60	0.85	43	51
C55/67	55	67	0.85	47	57
C60/75	60	75	0.85	51	64
C70/85	70	85	0.85	60	72
C80/95	80	95	0.85	68	81
C90/105	90	105	0.85	77	89
C100/115	100	115	0.85	85	98

*The relationship between *In-situ* and Potential Strength is based on good curing and full compaction conditions.

Criterion B
- mean of groups of n non-overlapping test results $\geq f_{cu} + k.s$
- each individual result $\geq f_{cu} - 4$
  f_{cu} is the characteristic *In-situ* Compressive Strength taken from Table 5.3
  n = the number of results, k = the related coefficient taken from Table 5.4
  s = the standard deviation of test results, but not less than 2 N/mm².

It is interesting to note the means of assessing core test results in the ACI code 301.

Core sampling and testing

Table 5.4 Coefficient k

Number of test results n	Coefficient k
9	1.67
10	1.62
11	1.58
12	1.55
13	1.52
14	1.50
≥15	1.48

Note: In order to achieve the same degree of confidence in assessments, the value of k is increased when the number of test results is low

This requires three cores to be taken from the suspect concrete and the average of the core tests to equal at least 85 per cent and with no single core result to fall below 75 per cent of the characteristic strength of the concrete under review. The South African Bureau of Standards SABS 0100-2:1992 (as amended 1994) applies the same criteria as the ACI but with different core strength relative to the specified concrete (see above).

If concrete is found to be non-compliant then the following actions are to be considered:

- strength requirements for the members
- performance of a full-scale load test
- strengthening the deficient part of the structure
- removal and replacement of the deficient part of the structure

5.4 Worked examples

5.4.1 Example 1

A concrete slab is found to have a characteristic strength of 28 N/mm^2 and density of 2350 kg/m^3 from on-going cube data. The prescribed grade is 30N. One truckload of concrete is suspected of non-compliance due to a colour variation. The location of the suspect batch is known but no cubes are available. Core sampling is requested by the Client on the suspect concrete. The minimum acceptable *In-situ* Cube Strength was set at 20.0 N/mm^2 (i.e. 30/1.5) during a planning meeting before cores were agreed to be taken.

Core data:
Cores 100 mm dia.   Area 7854 mm^2.
The following is a detailed calculation carried out on one core:

Prepared core length (ground)	= 104 mm
Length/diameter 104/100	= 1.04
One 10 mm bar in the core, 50 mm from one end	
Steel multiplication factor	$1 + 1.5[(10/100) \times (55/100)]$
	= 1.075

Voidage correction factor calc. $(2350 - 2280)/2350 - 500) \times 100\%$
$\qquad\qquad\qquad\qquad\qquad\qquad\qquad\qquad\qquad = 4.0\%$
$\qquad\qquad$visual$\qquad\qquad\qquad\qquad\qquad\qquad\quad = 3.0\%$
O/A Correction factor taken as 1.33 based on 3.5%

Curing correction factor taken as 1.0
Calculation:
Max. load at failure	= 145 kN
Strength (load/area)	= 18.46 N/mm^2
Estimated *In-situ* strength	= 18.46 × 2.3/(1.5 + 1/λ)
	= 17.25 N/mm^2
Correction for steel inclusion 17.25 × 1.075	= 18.54 N/mm^2
Minimum acceptable *In-situ* cube strength	= 30/1.5
	= 20.0 N/mm^2

The concrete can be taken as unacceptable in terms of *In-situ* Strength.
Average curing conditions were experienced on site, therefore:

Estimated Potential Strength	= Est. *In-situ* Str./0.77 (using CSTR No. 11)
	= 18.54/0.77
	= 24.08 N/mm^2 (24.0)
Excess voidage correction,	= 24.00 × 1.33
	= 31.92 N/mm^2 (32.00)
Estimated Potential Strength P	= 32.00 N/mm^2

The average of four cores gave an Estimated *In-situ* Cube Strength of 17.50 N/mm^2 and Estimated Potential Strength of 33.50 N/mm^2.

Conclusion

Since P, the Estimated Potential Strength, is greater than the characteristic strength of the concrete, the supply delivered to site has been shown to be in accordance with the specification. However, the *In-situ* Strength was below the minimum specified and the high excess voidage indicated that incomplete compaction was achieved. Note that it was the colour variation which brought about concern over this batch of concrete and that the concrete had not been compacted satisfactorily. It may be that the concrete was incapable of being compacted and that the colour variation was due to a different cement content than the rest of the slab.

It was recommended that the core samples be analysed for cement content and the concrete gang questioned to find out how easy the concrete was to compact. The delivery ticket should be examined to determine the time of batching and discharge on-site and any slump testing that may have been carried out. If a workability admixture was intended to be used, a check should be made at the batching plant to see if the records show it was included and consideration given to having the concrete analysed to determine if such an admixture is present in the suspect batch.

From the above, it can be seen that the core testing can apportion blame but a full picture of the circumstances at the time of placing needs to be established.

5.4.2 Example 2

An office block construction required a ground floor of Grade C30 concrete. The expected mean cube strength was 34 N/mm² from data held by the concrete supplier. Over a four-month period cube results averaged 35 N/mm², but over the next two months only a few cubes were taken and these gave an average strength of 22.5 N/mm². Ten of the cubes were below the characteristic strength of 30 N/mm² and two were below 20 N/mm² the minimum acceptable strength for the slab.

Cores were taken without planning or investigating the overall factors that had brought about the drop in cube strengths and an assessment made of the Potential Strength. Fifteen cores were cut vertically downwards in the slab, 100 mm diameter and 100 to 110 mm long after preparation. The Potential Density was agreed at 2350 kg/m³. The curing history was taken as 'normal' with the weather over the period noted and agreed as 'moderate'. The top 20 per cent of the samples was removed to omit any non-representative concrete. Average saturated density was found to be D_a = 2315 kg/m³ and the core strengths ranged between 15.5 to 24.5 N/mm².

Estimated Potential Strength

Visual estimates of excess voidage ranged between 0.5 and 2.0 per cent and averaged at 1.0 per cent.

Average estimates of excess voidage were determined

$$\frac{35 \times 100}{2350 - 500} = 2\%$$

The excess voidage value was taken as 1.5 and a correction factor of 1.13 was agreed for all cores. (Note that a general voidage factor may not always be appropriate and individual voidage correction factors may need to be applied to each core strength.)

The appropriate formula for Potential Strength is:

$$1.13 \times 3/(1.5 + 1/\lambda) \times \text{core strength.}$$

This yielded estimates of Potential Strength ranging from 32.0 to 36.5 N/mm².

Interpretation of results

1 All estimates of Potential Strength from cores exceeded the specified strength of 30 N/mm²
2 The Potential Strength estimates from cores were significantly different from the cube results over the two months in question.
3 The core densities were similar to the reported cube densities.

Conclusions

The test house carrying out the compression testing of the cubes was visited and it was noted that one set of platens were damaged and the faces not parallel. Fortunately, the cores were tested elsewhere. It was concluded that the concrete complied with the specification and that the low cube test results reported were caused by a set of defective platens.

5.5 Updating CSTR No. 11

CSTR No. 11 was first published in 1976 and an addendum added in 1987. Its procedures only relate to Portland cement concrete and therefore exclude pfa and ggbs concretes. Cement replacements and cementitious blends now form a significant sector in construction and so an update is needed to incorporate these. Further research was called for in CSTR No. 11 'so that a major improvement to the conversion of core strengths to Potential Strengths might be found.' Any improvement was thought to require:

- more information to be obtained on the difference between Actual (*In-situ*) and Potential Strength.
- account taken of thermal history.
- more accurate methods of assessing the reduction caused by voidage.
- more knowledge on the effect of age on the strength gain *in-situ*.

The 1987 Addendum called for:

- further evidence on small percentages of entrapped air.
- more specific data needed based upon pre-planned field investigations covering:
  - curing conditions
  - wide range of materials
  - mix proportions
  - varying construction conditions
  - types of structures
- more information on *In-situ* Cube Strength up to one year.
- fuller guidance to cover distribution of *In-situ* Strength within elements.
- research needed to be based on the properties in structures and not from test specimens.

5.5.1 Obtaining the required new data

In order to obtain the required data as set out above a research project was set up by the Concrete Society and funded by the DETR, BCA, and the ggbs and pfa industries. This extensive project included the following:

- Casting walls, slabs and blocks of concrete during both the summer and winter, a total of 96 structures.
- Cementitious blends including pfa, ggbs and limestone filler in Portland cement as well as straight Portland cement. Two addition levels providing two strength levels.
- Use of a crushed limestone and natural rounded gravel
- Batching without including an admixture so as to remove any uncertainty as to the contribution that an admixture might have on the data obtained.
- Structures cast using a concrete gang to build the shuttering and cast and compact the concrete as per a typical site operation.
- cubes taken and tested at 7 and 28 days, Core taken at 28, 42, 84 days and 1 and 3 years. Unfortunately the three-year cores were not tested.
- temperature-matched curing of the pfa and ggbs concretes.

The project reported data on 2500 core samples and 700 standard cubes. Further details of the project have been described by True (1999).

The project described above gave data that conflicted with the limited recommendations given in CSTR No. 11. Other data from Hong Kong reported by Watkins *et al.* (1996) and the South African Roads Board Research and Development Advisory Board (1990) confirmed the suspicion widely held on the general recommendations given in CSTR No. 11 and together these three documents, from three independent sources, have provided approximately 5000 core test results that collectively cast doubt on the findings of CSTR No. 11 and on the dogmatic way *In-situ* Strength and Potential Strength are related in prEN 13791 and omitted in prEN 206 as shown in Table 5.3.

5.5.2 Results of the new data

The data all show that concrete in structures continues to gain strength after the first 28 days which is assumed by CSTR No. 11 to be the time of maximum strength development of air cured Portland cement concrete (see Figure 5.1). Table 5.5 gives the ratios of *In-situ* Core Strength to 28-Day Cube Strength. This table presents the results without correction

Table 5.5 Ratios of *In-situ* core strengths to cube strength

Cement	Mix	Unit	Ratios of *In-situ* core strength to 28-day cube strength							
			28-day		42-day		84-day		365-day	
			Min	Max	Min	Max	Min	Max	Min	Max
PC	Lower	Block	0.75	0.95	0.80	0.95	0.85	1.05	0.95	1.10
	Strength	Slab	1.00	1.25	1.05	1.35	1.05	1.50	1.15	1.60
	Mixes	Wall	0.85	0.95	0.85	1.00	0.85	1.05	0.90	1.10
	Higher	Block	0.60	0.70	0.70	0.75	0.70	0.80	0.75	0.90
	Strength	Slab	0.85	1.00	0.95	1.10	0.95	1.15	1.05	1.25
	Mixes	Wall	0.85	0.95	0.80	0.95	0.85	1.00	0.95	1.05
P/FA-B	Lower	Block	0.85	1.10	0.85	1.20	0.95	1.35	1.45	1.65
	Strength	Slab	0.80	1.20	0.90	1.30	1.00	1.50	1.65	2.15
	Mixes	Wall	0.80	1.05	0.85	1.10	1.05	1.30	1.50	1.60
	Higher	Block	0.75	1.00	0.80	1.00	0.95	1.05	1.00	1.35
	Strength	Slab	0.75	0.95	1.00	1.15	1.15	1.25	1.20	1.70
	Mixes	Wall	0.80	1.00	0.90	1.05	1.05	1.25	1.35	1.40
P/B	Lower	Block	0.70	1.10	0.85	1.10	1.00	1.20	1.15	1.35
	Strength	Slab	0.65	1.15	0.85	1.30	1.20	1.40	1.35	1.70
	Mixes	Wall	0.70	1.00	0.85	1.15	1.05	1.25	1.15	1.35
	Higher	Block	0.80	1.05	0.95	1.20	0.90	1.25	1.00	1.55
	Strength	Slab	0.90	1.05	1.05	1.30	1.20	1.30	1.30	1.80
	Mixes	Wall	0.70	1.00	0.85	1.10	1.00	1.20	1.05	1.25
PLC	Lower	Block	0.80	0.95	0.85	1.00	0.95	1.10	1.00	1.15
	Strength	Slab	0.95	1.05	1.05	1.20	1.15	1.35	1.20	1.45
	Mixes	Wall	0.85	1.00	0.90	1.05	0.95	1.00	1.10	1.20
	Higher	Block	0.60	0.80	0.70	0.90	0.70	1.00	0.80	0.95
	Strength	Slab	0.80	0.90	0.95	1.00	1.00	1.05	1.05	1.20
	Mixes	Wall	0.80	0.95	0.90	1.00	0.90	1.05	1.00	1.15

Notes: (1) PC is Portland cement, P/FA-B is a 30% PFA: 70% PC blend, P/B is a 50% ggbs: 50% PC blend, PLC is a Portland Limestone Cement.
(2) Blocks were 1.5 m cubes, slabs $2.0 \times 2.0 \times 0.2$ m thick and walls $3 \times 2 \times 0.3$ m thick.
(3) The above ratios do not include an orientation factor.
(4) Concrete grades were nominally C30 and C50.

for orientation and gives the results as ranges which shows that the general 0.77 factor used in CSTR No. 11 and the 0.85 factor in prEN 13791 are both incorrect. It is to be remembered that if core sampling and testing is carried out because cube data appears inconclusive, the core samples will be older than 28 days and Table 5.5 shows a significant change in the ratio with age, especially for the pfa and ggbs concretes.

The results from the Hong Kong trials reported by Watkins *et al.* (1996) both showed that *In-situ* Strength was very similar to Cube Strength and core density was very close to cube density. Data on density from the Concrete Society project discussed above indicated that voidage increased as the unit volume increased with walls and slabs having a 1 per cent and blocks 2.5 per cent excess voidage respectively.

Data from the two South African studies concluded with an extensive statistical analysis and showed it was impossible to derive a general reliable strength ratio. The results were reported to defy rational analysis and the PCI laboratories discontinued providing potential strength estimates.

It is therefore important that all interested parties to any core sampling and test programme first, agree the need for the test and second, agree the sampling locations and means of interpretation of the data **before** any cores are taken. If cores are needed to determine the *In-situ* Strength, the above agreements can usually be resolved by those knowledgeable in the subject. When Potential Strength estimates are planned, it can be seen from the preceding sections that the interpretation is not straightforward and requires careful consideration with the realization that any estimate will be liable to considerable inaccuracy.

5.5.3 Other considerations

It is perhaps useful to the reader who may be about to carry out core sampling to consider further the affects of curing, orientation factoring and allowance for excess voidage. These three can have significant effect on the outcome of a Potential Strength assessment.

Curing

In-situ curing affect on a structure is considered by some to be the most uncertain aspect when deriving Potential Strength estimates. Fortunately, guidance is found in the Concrete Society Study Group report (Cather, 1992) and the CIRIA Funders Report/CP/38 (1996) which both refer to a curing affected zone (CAZ) beyond which the external environment has virtually no effect on internal humidity. This depth is commonly taken as 50 mm for normal density concrete. Core sampling needs to consider the location of the test pieces relative to the CAZ.

Orientation factor

CSTR No. 11 applies by formulae an orientation factor of 8.7 per cent and suggests that vertically cut cores are stronger, in all concretes, to horizontal ones. Only limited data is provided to substantiate this. Others, including Petersons (1971) and Watkins *et al.* (1996) have published data that suggests this is too high a correction factor. Indeed, Peterson states that the direction cores are drilled is of no importance and the data from the Concrete Society project reported by Truc (1999) would suggest a factor of 4 per cent as the mode of a range of 0.57 to 1.56 and any value taken is therefore affected significantly by other factors. If concrete is not prone to excessive segregation and the aggregate is typically free of flaky and elongated particles in any significant quantities, correction for core orientation is unlikely to be needed. However, judgement on the matter is best made by examining the surface of the cores, noting the orientation, voidage shape and accumulation relative to the aggregate.

Voidage

If curing effects on a structure are uncertain, they are small when compared with the factors used to correct for voidage (see Table 5.2).

A typical excess voidage may be 2.5 per cent and this suggests a strength multiplying factor of 1.23 to be applied to the core strength. Voidage is assessed by two methods in CSTR No.11, one based on density measured by weighing in air and water, the other on relating the appearance of the cores to photographs of stated known voidage. Core testing, especially when carried out because doubt is cast on standard cube strengths, may also consider the cube density as unrepresentative. The baseline for assessing excess voidage by density is the cube density.

Even if excess voidage can be agreed, a suitable strength correction factor may not be easily selected from the above table. Warren (1973) published data which is presented in Figure 5.2 that shows a considerable influence on the voidge correction factor brought about by both concrete grade and aggregate type. CSTR No. 11 has just taken the mean of the lower end of these results to compile their correction factors.

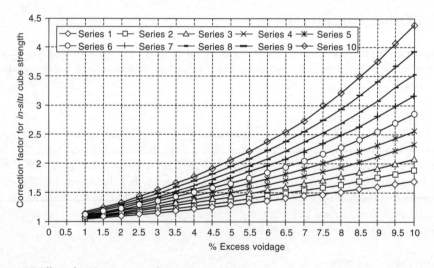

Figure 5.2 Effect of voids on measured strength according to Warren.

One objective means of assessing voidage is under way by which each core sample is scanned and the image used to assess voidage. Control images of known voidage are used to calibrate software and derive a true voidage from the sample. Once the cores have been crushed, a strength/voidage or density relationship can be derived and applied to correct the core strength. This work is now being progressed by the author.

5.5.4 The effect of voidage and potential density on potential strength estimates

Potential Strength estimates require voidage and voidage correction factors to be applied to the *In-situ* Strength obtained from core test results. The visual assessment method is subjective and has insufficient examples in CSTR No. 11 to enable an accurate comparative assessment.

Density assessment by weighing in air and water has poor reproducibility, as assessed from cube test data unpublished by the BCA, who reported 35 kg/m^3 which relates to a voidage of approximately 2 per cent and a corresponding strength factor of 18 per cent.

CSTR No. 11 assumes a single relationship between voidage and strength but others have suggested aggregate type, water/cement ratio and/or concrete strength may affect the relationship. When one considers a typical voidage factor of 2.5 per cent with an associated reproducibility of 2 per cent the range of correction applied to the *In-situ* Strength becomes significant and likely to cast doubt over the usefulness of the assessment i.e. 23 per cent ± 18 per cent. Price (1995) concluded that it is a questionable procedure and it may be time to consider omitting it from the revised CSTR No. 11. Indeed, if emphasis is placed on correctly sampling, making, curing and testing cubes, at a suitable frequency, there would be no need to derive Potential Strength estimates.

5.5.5 Interpretation options

Core strengths are usually modified and interpreted to derive estimates of either *In-situ* Strength or Potential Strength. Which option is taken, and how the results are interpreted must be decided before core sampling and testing. *In-situ* Strength interpretation is relatively straightforward and unlikely to provide contentious results.

A procedure for deriving *In-situ* Strength estimates has been detailed in section 5.3.8 and relates to converting a core strength result to an equivalent cube result, accounting for any steel inclusion and orientation. Perhaps consideration needs to be applied to the orientation factor in light of the findings discussed in section 5.4.3. It is also worth remembering that an estimate of *In-situ* Strength requires only one core result but that will provide an estimate of ±12 per cent of the true value. The accuracy can be improved to $\pm 12/\sqrt{n}\,\%$, where n is the number of core results included. Consideration must be given at the planning stage to the accuracy required which will assist interpretation of the results.

Derivation of Potential Strength estimates is a contentious exercise. CSTR No. 11 is still a current report and its application to derive Potential Strength estimates of normal weight, Portland cement concrete is open to those wishing to use the document. However, while awaiting an update of CSTR No. 11 and in light of the findings of the Concrete Society cores project, one should consider carefully the formulae and factors presented in CSTR No. 11.

One approach suggested by Troy and True (1994) is to derive Potential Strength estimates from *In-situ* Strength results. This is largely supported by Price (1995) who also favours *in-situ* testing rather than recourse to an increasingly flawed procedure such as the estimation of Potential Strength.

Levitt, in a submission to the Concrete Society Materials Group, suggested the approach of modifying *In-situ* Strength results by a simplified range comparison based on using a master table such as given in Table 5.5. If one considers a C40 concrete using P/FA-B concrete that gave typical 28-day cube results of 38–50 N/mm^2 and cores were taken at one year which gave *In-situ* Strengths of 40 to 52.5 N/mm^2, the means of checking for compliance at 28 days using the core results would be:

Factor range for P/FA-B, slab at 1 year = low strength 1.65 to 2.15
high strength 1.20 to 1.70
take 1.43 to 1.93

Calculation of the estimated *In-situ* Cube Strength at 28 days:
40/1.93 to 52.5/1.43
20.5 to 36.5 N/mm^2

Grade of concrete is therefore:
20.5 + 0.05(36.5 − 20.5)
21.3 N/mm^2

That is C21 and the concrete is shown to be non-compliant with the 28-day strength requirement of 40 − 4 N/mm^2, i.e. 36 N/mm^2.

Until a more thorough review of CSTR No. 11 has been undertaken and the document either updated or completely rewritten, a pragmatic approach may be to follow the acceptance criteria suggested by Addis (1992) and included after modification in the South African Standard SABS 0100-2:1992 (As amended 1994). This states:

14.4.3 Acceptance of concrete on the basis of core strengths

14.4.3.1 If the average core strength is at least 80% of the specified strength, and if no single core strength is less than 70% of the specified strength, the concrete shall be accepted.

14.4.3.2 If the concrete in a certain area fails to comply with 14.4.3.1 because a single core result falls below 70% of the specified strength, a further set of three cores may be taken from the same area, to determine the extent of deficient concrete. If the new set of three cores complies with the requirements of 14.4.3.1, the area represented by this second set of cores shall be considered acceptable. If the new set of cores fails to comply with the requirements of 14.4.3.1, 14.4.3.3 applies

14.4.3.3 If the core strength does not meet the acceptance criteria of 14.4.3.1 or 14.4.3.2, the following should be considered.
 (a) strength requirements for the member(s);
 (b) performance of a full-scale load test;
 (c) strengthening the deficient part of the structure; and
 (d) removal and replacement of the deficient part of the structure.

The latest draft proposal being considered by the Concrete Society for updating CSTR No. 11 is based on the following interpretation of core strengths:

- core strengths used to derive *In-situ* Cube Strength. 100 mm diameter cores, 100 to 120 mm long, converted directly to derive an Estimated Potential Strength.
- removal of the equations in CSTR No. 11 for deriving Potential Strength.

- removal of universal orientation factors.
- applying factors to the derived Estimated Potential Strength. It has been assumed by the Working Party that all the relevant factors are included in this list. However, the value or range of values to be attributed by applying each of these factors is still under debate. Indeed there is considerable work to be undertaken to resolve all the uncertainties related to some of these factors. The factors are:

F_1 length to diameter
F_2 amount, diameter and location of reinforcement
F_3 orientation factor
F_4 core age at time of test
F_5 correction for compaction and excess voidage
F_6 curing regime – summer, winter
F_7 type of cement
F_8 air-entrainment
F_9 type and size of structural element
F_{10} aggregate type, lightweight, rounded, crushed etc.

References

Addis B.J. (1992) A rational approach to the assessment of concrete strength on the basis of core tests. *Concrete Beton*, November.
American Concrete Institute. Specification for structural concrete for buildings. ACI 301-84, Chapter 17, clause 17.3.2.3.
ASTM C.42 Standard Method of obtaining and testing cores and sawed beams of concrete.
Bowman, S.A.W. (1980) Discussion on ref. 71. *Concrete Research,* **32**, No. 111, June, p. 124.
BS 6089:1981 Guide to assessment of concrete strength in existing structures. (To be replaced by prEN 13791 Assessment of concrete compressive strength in structures or in structural elements.)
BS EN 12504-1:2000 Testing concrete in structures Part 1: Cored specimens – Taking, examining and testing in compression. (Replaces BS 1881-120:1983 Testing concrete Part 120: Method for determination of compressive strength of concrete cores.)
BS EN 12350-1:2000 Testing fresh concrete – Part 1: Sampling.
BS EN 12390-2: 2000 Testing hardened concrete – Part 2: Making and curing specimens for strength tests.
BS EN 12390-3: 2002 Testing hardened concrete – Part 3: Compressive strength of test specimens.
BS EN 12390-4: 2000 Testing hardened concrete – Part 4: Compressive strength – Specification of compression testing machines.
British Cement Association. Precision of BS 1881 concrete tests. Part 1: Data and analysis. April 1988. Unpublished.
Bungey, J.H. (1979) Determining concrete strength by using small diameter cores. *Concrete Research*, **31**, No. 107, June, 91–98.
Cather, R. (1992) Concrete Society Study Group Report. How to get better curing. *Concrete*. **26**, No. 5. Sept/Oct, 22–25.
CIRIA, (1996) Funders Report/CP/38. On-site curing – Influence on the durability of concrete: a review.
Concrete Society (1976) Technical Report No. 11 (and Addendum (1987)) Concrete core testing for strength. Report of a Concrete Society Working Party, May.
Concrete Society (1997) Technical Report, Core strength of *In-situ* Concrete. Results of a DETR and British Cement Association funded project under the Partners in Technology scheme managed by the Construction Sponsorship Directorate.

Levitt, M. Submission to the Concrete Society Materials Group. Possible alternative approach to a revision of CSTR 11.

Neville, A.M. (1995) *Properties of Concrete*, Longman, Harlow, 612.

Petersons, N. (1971) Recommendations for estimation of quality of concrete in finished structures. *Materiaux et Constructions*, **4**, No. 24.

Price, W. (1995) Potential strength – cores for concern. *Concrete.* **29**, No. 5. Sept/Oct, 51–52.

Robins, P.J. (1980) The point-load strength test for cores. *Concrete Research*, **32**, No. 111, June, 101–111.

South African Roads Board Research and Development Advisory Committee. *Project Reports PR 89, Vol 1, 1990; PR 101, Vol 2, 1992*. Strength parameters for concrete cores. Research carried out by the Portland Cement Institute.

South African Standard SABS 0100-2:1992 Code of practice. The structural use of concrete Part 2: Materials and execution of work (Including Amendment No. 1, Sept. 1994).

Troy, J.F. and True, G.F. (1994) A case of cores and defect. *Concrete.* **28**, No. 4, July/August pp 24–25.

True, G.F. (1999) Are we closer to interpreting concrete core strengths? *Concrete*, **33**, No. 2, 29–32.

Warren, P.A. (1973) The effects of voids on measured strength. Egham, RMC Technical Services Ltd, Technical Note No. 73.

Watkins, R.A.M. *et al.* (1996) A comparison between cube strength and *in-situ* concrete strength development. In *Proceedings of the Instn. of Civ. Engr. Structs. and Bldgs*, May 1996, 116, 138–153.

6

Diagnosis, inspection, testing and repair of reinforced concrete structures

Michael Grantham

6.1 Introduction

Anyone involved in the maintenance of reinforced concrete structures cannot be unaware of the substantial increase in the amount of repair work now being carried out. This may merely reflect the fact that a large number of reinforced concrete structures are now coming of age although certain changes in materials have been identified over the past twenty years or so which might be partially responsible in some cases.

For example, in the early 1970s it was believed that there were no deposits of alkali-reactive aggregates in the UK and yet within a few years alkali aggregate reaction became a subject of major interest as more and more cases came to light. This upsurge in ASR problems may be partially traceable to an increase in cement alkali content in the preceding years. Furthermore, increases in the early age strength of cement may have meant that specified minimum strengths were being achieved with less cement than had previously been used with an appropriate effect on subsequent durability.

In the light of such phenomena it is clear that concrete must be regarded as a complex mixture of materials each component of which may, in itself, vary or be affected by environmental factors.

From experience it is true to say that 90 per cent of the problems that will be experienced in concrete repair will involve steel reinforcement corrosion as a primary problem. For the most part this will have been caused entirely by simple carbonation/low cover and/or the presence of chloride salts either from calcium chloride used as an accelerator, or from de-icing salt.

However, in some cases, some other more subtle defect may be present, such as a shrinkable aggregate, alkali–silica reaction, freeze–thaw damage, sulfate attack, structural cracks or a whole variety of other possibilities. It must be borne in mind that many of these phenomena may reveal themselves first of all in areas of low cover and carbonated concrete, perhaps because microcracking from one or other of these causes has permitted carbonation to advance more rapidly than might otherwise have been the case. In such cases, it is all too easy to look at the effects of the problem, i.e. steel corrosion, and attribute this as the cause. Attempting to repair concrete affected by such problems may simply mean that the problem recurs in a relatively short space of time.

6.2 What is concrete?

Before considering some of the defects that can appear in reinforced concrete it is worth pausing briefly to examine the components of the concrete and some of the potential problems that can occur as a result of the materials themselves.

6.2.1 Cement

Cement
The significance of the alkali content in relation to possible alkali–silica reaction needs to be considered. Have appropriate precautions been taken to avoid thermal cracking in high cement content deep-section members? Has the concrete been adequately cured and appropriate attention been given to the avoidance (as far as possible) of shrinkage cracking?

Sulfate-resisting cement
Where chloride salts are present, sulfate-resisting cement may show an increased susceptibility to reinforcement corrosion.

High alumina cement
The converted form of HAC may show a reduction in strength and become more susceptible to certain forms of chemical attack. More recently there has been increased concern over the occurrence of carbonation in HAC which has been responsible for problems with reinforcement corrosion on top of the problems due to loss of strength from to conversion.

6.2.2 Water

A good general rule is that if the water is drinkable it should be suitable for concreting purposes and the vast majority of water employed for concrete will have been drawn from the mains. However, circumstances may dictate the use of other sources and (very occasionally) problems may be encountered.

Consideration should be given to the presence of dissolved salts, organic matter, sugar, lead, etc., which can retard the setting of the cement. Much more frequently the effects of an excessive or inadequate water content will be encountered; resulting respectively in a porous concrete or a concrete which is difficult to compact.

6.2.3 Aggregate

The aggregates employed in the UK and Ireland reflect a rich geological record and cover a vast range of different geological types of material. Typical problems encountered may include:

(a) The presence of impurities such as organic matter, sulfates, chlorides or sodium or potassium salts.
(b) Some aggregates may have poor physical properties such as aggregate crushing value or flakiness. They may be shrinkable, expandable, porous or frost susceptible.
(c) The presence of constituents which are susceptible to alkali reaction.
(d) Poorly graded material.

6.2.4 Steel

The reinforcement may consist of one or more of a variety of types; mild steel, high-yield steel, weathering steel, ferritic, stainless or galvanized steels. It is true to say that, of itself, the steel is rarely responsible for problems with reinforced concrete. Corrosion of the steel, where it is put or whether it is put there at all, on the other hand, are much more frequently encountered!

6.2.5 Admixtures

Until recent years admixtures were rarely used in the UK (except possibly air-entraining admixtures) and again problems resulting from admixtures are rare. Problems which may be encountered include:

(a) *Air-entraining admixtures* An overdose will produce a reduction in strength and may aggravate freeze–thaw damage. Insufficient may not be effective in preventing frost damage.
(b) *Plasticizers* A very large overdose can retard the setting of the concrete.
(c) *Chlorides* Older structures in which calcium chloride was employed may be suffering from reinforcement corrosion.
(d) *Retarders* These may be overdosed.
(e) *Polymeric bonding agents* Unless correctly used, these can make very efficient de-bonders!

6.3 Recognizing concrete defects

6.3.1 Structural failure

Actual structural failure, or even structural cracking, is only rarely encountered but it is important to differentiate between cracking from structural and other causes. Such an assessment should only be carried out by a structural engineer, but an initial inspection by a materials engineer may highlight other (much more likely) causes of cracking. The materials engineer will know when the advice of a structural engineer is required. Figure 6.1 illustrates structural cracks at a column beam connection. If examined carefully, the column behind shows no evidence of cracking, but in this case the beam is supported on a brick wall. The wall was not intended to be loadbearing, but was clearly taking some of the stress off the column

Figure 6.1 Structural cracks at a column beam connection.

6.3.2 Corrosion of steel

Steel reinforcement is normally chemically protected from corrosion by the alkaline nature of the concrete. If this alkalinity is lost through carbonation or if chlorides are present which can break down this immunity, then corrosion can occur. Obviously, when cover is low, the onset of corrosion will be sooner.

The type of corrosion which occurs varies. Carbonation-induced corrosion tends to affect large areas of the bar causing a gradual loss of section over a relatively wide area. The corrosion problem is obvious before serious damage can be done because the concrete cover will spall. With chlorides, a different mechanism often occurs causing very localized severe loss of section. This can occur without disruption of the cover concrete and almost total corrosion of section can occur before problems become apparent at the surface. Where pre-stressed steel is used, catastrophic failures have occurred with no prior warning in one case on a structure which had been load tested shortly before the failure occurred.

Figure 6.2 illustrates the problems which occur when the original cover provided is inadequate. Correct use of a covermeter at the construction stage can avoid these easily preventable problems.

Figure 6.2 Inadequate cover of steel reinforcement.

6.3.3 Alkali–silica reaction

Alkali–silica reaction can occur in concretes made with aggregates containing reactive silica, provided there is a sufficient supply of alkali (usually provided by the cement) and a supply of moisture. The reaction product is a hygroscopic gel which takes up water and swells. This may create internal stresses sufficient to crack the concrete.

One of the most frequently found aggregates in affected concrete is chert. This is a common constituent of many gravel aggregates, but a number of other geological types may be reactive, such as strained quartz in sands and some quartzites. Some Irish aggregates, notably greywackes, have been found to be susceptible to ASR. These tend to be quite slow reacting and damage can take 20–50 years to become serious.

Figure 6.3 illustrates cracking to the Beauharnois Dam on the St Lawrence Seaway. The expansion caused by the ASR on this structure caused it to grow several centimetres in length and distorted the turbines providing hydroelectric power for the region.

6.3.4 Freeze–thaw damage

Concrete of inadequate durability, if subjected to a wet environment and freezing, can be disrupted by freeze–thaw attack. Water enclosed in the pores of the wet concrete will

Figure 6.3 Cracking caused by alkali–silica reaction.

expand on freezing and the high internal stresses so created can disrupt the surface. The effects are intensified by subsequent freeze/thaw action as minute cracks develop which, in turn, become filled with water. This progressive damage causes layers of laminar cracking parallel to the surface – a classic indication of freeze–thaw damage.

6.3.5 Shrinkable aggregates

Some, mostly igneous, aggregates can contain inclusions of weathered material in the form of clay minerals. These minerals, in common with the clays encountered in the ground, swell in the presence of moisture and shrink as they dry out. They can cause excessive drying shrinkage of the concrete and can cause a random crack pattern not unlike that encountered with ASR. The problem was first identified in Scotland, where it is quite common, but has been observed in the North East of England, Hertfordshire, Wales and Cornwall. The cracking can pose potential structural problems but is more likely to cause loss of durability and is frequently associated with freeze–thaw damage. Figure 6.4 illustrates a severe example accompanied by pop-outs where frost has attacked the porous aggregate.

6.3.6 Chemical attack

Sulfate attack
Concrete buried in soils or groundwater containing high levels of sulfate salts, particularly in the form of sodium, potassium or magnesium salts, may be subjected to sulfate attack

Figure 6.4 Damage from shrinkable aggregates accompanied by frost attack.

under damp conditions. An expansive reaction occurs between the sulfates and the C_3A phase to form calcium sulfoaluminate (ettringite) with consequent disruption to the matrix. Past experience has shown that true sulfate attack is rare in concrete, only occurring with very low cement content concretes, with less than about 300 kg/m^3 of cement. As a guide, levels of sulfate above about 4 per cent of cement (expressed as SO_3) may indicate the possibility of sulfate attack, provided sufficient moisture is present. Sulfate attack requires prolonged exposure to damp conditions. However, there has been recent concern with another form of sulfate attack, as follows:

Thaumasite attack – a form of sulfate attack

This hit the news in 1998 when the foundations to a number of bridges on the M5 motorway in the UK were found to be suffering from serious erosion and crumbling of the outer part of the concrete in the foundations. The problem was diagnosed as being due to an unusual form of sulfate attack, known as thaumasite attack. For the problem to occur, a number of factors have to be present:

A source of sulfate
Water (usually plenty of moisture)
A source of carbonate (as limestone aggregate, or filler, or possibly even as fill)
Low temperatures (<15°C) although cases have occurred at higher temperatures.

The combination of these factors can cause an unusual reaction between the cement, the lime and the sulfate, to form *thaumasite*, a sulfate mineral. The effect is to cause serious damage and softening of the exposed outer surface of the concrete (assuming an external source of sulfate). Figures 6.5 and 6.6 show one of the affected foundations and also a damaged road in North Texas which showed serious heaving of the road surface over a gypsum rich soil, when a lime soil stabilizer was used under the roadway.

It should be noted that sulfate-resisting cement has not proved to be any more resistant to normal Portland cement in resisting this type of attack.

Figure 6.5 Foundation damaged by thaumasite attack.

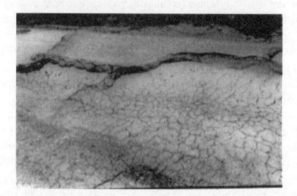

Figure 6.6 Road in north Texas heaving due to thaumasite/ettringite attack.

Acid attack

Typified by 'raised' aggregate or in extreme cases disintegration. This can sometimes be difficult to diagnose since the acid is neutralized by the cement paste and may be washed away. The most common sources are spillage from acid tanks, acidic groundwater and oxidation of sewage effluents.

Other contaminants

In addition to those specifically mentioned above, many potentially corrosive substances may come into contact with reinforced concrete. The extent of their effect may depend on the type of cement used. Examples of such materials are certain alkalis, beer and wine (carbonic acid, lactic acid and acetic acid), vegetable and fish oils, milk (lactic acid), lime, sugar and sulfides.

6.3.7 Fire damage

Effects on concrete due to fire

Damage to concrete attributable to fire has been summarized in Technical Report No. 15 of the Concrete Society. Three principal types of alteration are usually responsible (Figure 6.7):

1. **Cracking and microcracking in the surface zone** This is usually sub-parallel to the external surface and leads to flaking and breaking away of surface layers. Cracks also commonly develop along aggregate surfaces – presumably reflecting the differences in coefficient of linear expansion between cement paste and aggregate. Larger cracks can occur, particularly where reinforcement is affected by the increase in temperature.
2. **Alteration of the phases in aggregate and paste** The main changes occurring in aggregate and paste relate to oxidation and dehydration. Loss of moisture can be rapid and probably influences crack development. The paste generally changes colour and various colour zones can develop. A change from buff or cream to pink tends to occur at about 300°C and from pink to whitish grey at about 600°C. Certain types of aggregate also show these colour changes which can sometimes be seen within individual aggregate particles. The change from a normal to light paste colour to pink is most marked. It occurs in some limestones and some siliceous rocks – particularly certain flints and chert. It can also be found in the feldspars of some granites and in various other rock types. It is likely that the temperature at which the colour changes occur varies somewhat from concrete to concrete and if accurate temperature profiles are required, some calibrating experiments need to be carried out.
3. **Dehydration of the cement hydrates** This can take place within the concrete at temperatures a little above 100°C. It is often possible to detect a broad zone of slightly porous light buff paste which represents the dehydrated zone between 100°C and 300°C. It can be important, in reinforced or pre-stressed concrete, to establish the maximum depth of the 100°C isotherm

Figure 6.7

Changes in fire damaged concrete

<300°C	Boundary cracking alone
250–300°C	Aggregate colour changes to pink to red
300°C	Paste develops a brown or pinkish colour
300–500°C	Serious cracking in paste
400–450°C	Portlandite converts to lime
500°C	Change to anisotropic paste
500–600°C	Paste changes from red or purple to grey
573°C	Quartz gives a rapid expansion resulting from a phase change from alpha to beta quartz
600–750°C	Limestone particles become chalky white
900°C	Carbonates start to shrink
950–1000°C	Paste changes from grey to buff

Changes in aggregate

250–300°C	Aggregate colour changes to pink to red
573°C	Quartz gives a rapid expansion resulting from a phase change from alpha to beta quartz
600–750°C	Limestone particles become chalky white
900°C	Carbonates start to shrink

Changes in the paste

300°C	Paste develops a brown or pinkish colour
400–450°C	Portlandite converts to lime
500–600°C	Paste changes from red or purple to grey
950–1000°C	Paste changes from grey to buff

Cracking

<300°C	Boundary cracking alone
300–500°C	Serious cracking in paste
500°C	Change to anisotropic paste

The depth of colour change is often a good guide to the overall depth of damage. This may often be no more than a few millimetres for short-duration fires, even if high temperatures are reached at the surface. As a general rule, if no spalling has occurred, no damage is likely to have occurred to the steel. (This does not apply to pre-stressed concrete.)

6.3.8 Poor-quality construction

During construction, lack of attention to proper quality control can produce concrete which may be inferior in both durability and strength to that assumed by the designer. Particular factors in this respect are compaction, curing conditions, low cement content, incorrect aggregate grading, incorrect water cement ratio and inadequate cover to reinforcement (Figure 6.8).

Figure 6.8 Poor quality construction.

6.3.9 Plastic cracking

Cracks due to this phenomenon appear within the first two hours after placing and are of two distinct types:

- *Plastic settlement cracks*
  Typically found in columns, deep beams or walls. The problem tends to occur with high water/cement ratio concretes which have suffered from bleeding. The concrete literally 'hangs up' on the steel slumping between it, with cracks forming over the line of the steel. Caught early enough, re-vibration of the concrete can repair the damage while the concrete is still plastic. Figure 6.9 shows a core taken through a plastic settlement crack in a car park. The damage that the easy pathway for chlorides has caused can easily be seen.
- *Plastic shrinkage cracks*
  More common in flat exposed slabs. This can occur anywhere where the rate of loss of moisture due to evaporation exceeds the rate of bleeding. Not surprisingly it is more of a problem in hot, dry climates, but can easily occur on hot days in flat slabs especially where adequate attention to protection and curing has not been given.

Note: Neither of these types of crack should be confused with drying shrinkage cracks which only occur after a considerable time.

6.3.10 Thermal cracking and delayed ettringite formation

Thermal cracking of concrete can occur in large pours. Typically concrete can gain in temperature about 14°C per 100 kg of cement in a cubic metre of concrete. In large pours

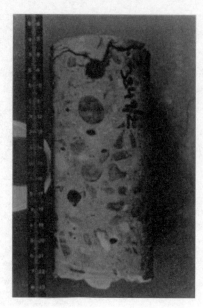

Figure 6.9 Plastic settlement crack.

this sets up a thermal gradient, with the outer part of the concrete cooling more rapidly than the core. This puts the outer skin in tension, and small cracks form. With the addition of subsequent drying shrinkage, the cracks can become quite large.

During the hydration process, it is quite normal for ettringite (calcium sulfoaluminate) to form inside the concrete. This mineral is normally associated with sulfate attack, but in the context of a setting concrete is quite normal. Any expansion resulting from its formation is taken up in the still plastic concrete.

However, if the temperature of the concrete exceeds about 60°C then formation of the ettringite can be delayed until after the concrete has hardened. In this situation, if a source of moisture is present, then the concrete can suffer from quite severe expansion and cracking.

There is a correlation with the alkali content of the cement, too. The higher the alkali content, the lower the temperature at which DEF can occur (Figures 6.10–6.13).

6.4 Investigation of reinforced concrete deterioration

Experience has shown that a number of testing methods are of proven value in determining the extent of deterioration of a concrete structure and in identifying those areas where remedial measures are necessary. While the list of tests given below is not exhaustive, it does include most of the common tests as well as one or two lesser known techniques.

6.4.1 The two-stage approach

Any investigation can conveniently be split into two stages:

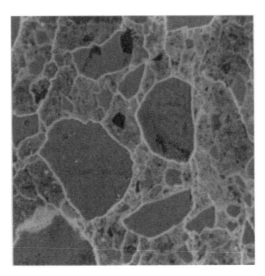

Figure 6.10 Thin section showing typical cracks around aggregate particles due to DEF. In this case, the *paste* expands, rather than the aggregate, as occurs with ASR.

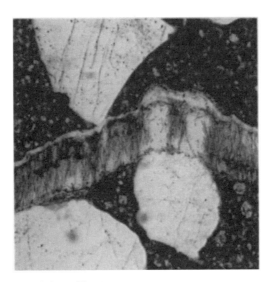

Figure 6.11 Ettringite in a crack formed by DEF.

Stage 1 An initial survey to identify the cause of the problems.
Stage 2 An extension of the stage 1 survey, perhaps using a limited number of techniques to identify the extent of the defects revealed by stage 1.

The advantages of such an approach are clear. In the stage 1 survey, work can be carried out on selected areas showing typical defects but choosing these, as far as possible, from areas with simple access, i.e. ground level, roof level, from balconies, etc. Occasionally, a lightweight scaffold tower or an electrically powered hydraulic lift can be used to advantage. One or more areas apparently free from defect would also be examined in this

Table 6.1 Classification of intrinsic cracks

Type of cracking	Letter (See Figure 6.13)	Subdivision	Most common location	Primary cause (excluding restraint)	Secondary causes/factors	Remedy (assuming basic redesign is impossible) In all cases reduce restraint	Time of appearance
Plastic settlement	A	Over Reinforcement	Deep sections	Excess bleeding	Rapid early drying conditions	Reduce bleeding (air entrainment) or re-vibrate	10 minutes to 3 hours
	B	Arching	Top of columns				
	C	Change of depth	Trough and waffle slabs				
Plastic shrinkage	D	Diagonal	Roads and slabs	Rapid early drying	Low rate of bleeding	Improve early curing	30 minutes to 6 hours
	E	Random	Reinforced concrete slabs				
	F	Over-reinforcement	Reinforced concrete slabs	Ditto plus steel near surface			
Early thermal contraction	G	External restraint	Thick walls	Excess heat generation	Rapid cooling	Reduce heat and/or insulate	One day to two or three weeks
	H	Internal restraint	Thick slabs	Excess temperature gradients			
Long-term drying shrinkage	I		Thin slabs (and walls)	Inefficient joints	Excess shrinkage, inefficient curing	Reduce water content, improve curing	Several weeks or months
Crazing	J	Against formwork	Fair-faced concrete	Impermeable formwork	Rich mixes, poor curing	Improve curing and finishing	One to seven days, sometimes much later
	K	Floated concrete	Slabs	Over-trowelling			
Corrosion of reinforcement	L	Natural	Columns and beams	Lack of cover	Poor-quality concrete	Eliminate causes listed	More than two years
	M	Calcium chloride	Pre-cast concrete	Excess calcium chloride			
Alkali–silica reaction	N		(Damp locations)	Reactive aggregate plus high-alkali cement		Eliminate causes listed	More than 5 years

Note: Figures 6.10–6.13 and this table are reproduced from Concrete Society Technical Report No. 22, *Non Structural Cracks in Concrete*.

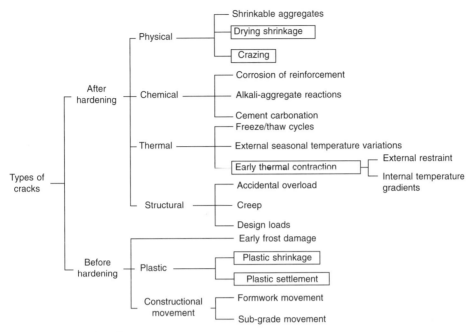

Figure 6.12 Types of crack.

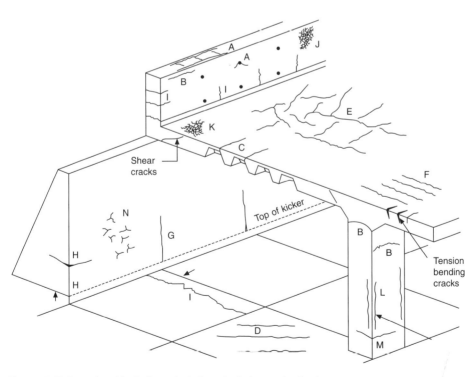

Figure 6.13 Examples of intrinsic cracks in hypothetical concrete structure.

initial survey as it is frequently found that, by comparing good areas with bad, the reason for the problems emerges by simple comparison.

In stage 2, once the defects have been identified, it is often necessary to quantify the extent of the problems. This may be as simple as carrying out a covermeter survey over the whole structure, where low cover has been identified as the problem, to the application of one or more of the other techniques described below.

6.4.2 Visual survey

After collecting as much background data as possible, any testing programme should begin with a thorough visual survey of the structure. This may conveniently be recorded on a developed elevation giving particular attention to the following defects:

- Cracks or crazing
- Spalling
- Corrosion of steel and rust staining
- Hollow surfaces
- Honeycombing due to poor compaction or grout loss
- Varying colour or texture
- Areas in which remedial finishing work has already been carried out
- External contamination or surface deposits
- Wet or damp surfaces

Throughout the course of any investigation colour photographs should be taken of points of particular interest.

6.4.3 Covermeter survey

Adequate cover to the steel reinforcement in a structure is important to ensure that the steel is maintained at a sufficient depth into the concrete so as to be well away from the effects of carbonation or from aggressive chemicals. However, excessively deep cover has its own problems; crack widths may be increased and the lever-arm decreased.

All covermeters are electromagnetic in operation. Electric currents in a coil winding in the search head generate a magnetic field which propagates through the concrete and will interact with any buried metal present, such as reinforcing steel. The interaction will be due to either or both of two physical properties of the steel: its magnetic permeability and its electrical conductivity. The interaction causes a secondary magnetic field to propagate back to the head where it is detected by a second coil or, in some instruments, by modifying the primary field. The signal received will increase with increasing bar size and decrease with increasing bar distance (cover). By making certain assumptions about the bar, and specifically by assuming that only one bar is present within the primary magnetic field, the instrument can be calibrated to convert signal strength to distance and hence to indicate the depth of cover.

If there is more than one bar (or even scaffolding!) within the range of the primary field the instrument will receive a greater signal and indicate a shallower cover than the true cover. The skilled operator will always carefully map out the position and orientation of the steel, breaking out some steel if necessary, to ensure that accurate results are obtained.

Some manufacturers claim that the size of the reinforcing bar may be determined by the use of spacer blocks and some inbuilt mathematical processing, or with some of the more sophisticated machines by internal data processing only. Such methods work satisfactorily only where a single bar is present within the range of the search head, but can be reasonably accurate with some of the more modern devices.

British Standard 1881: Part 204: 1988 requires that when measuring cover to a single bar under laboratory conditions, the error in indicated cover should be no more than plus or minus 5 per cent or 2 mm whichever is the greater. For site conditions, an average accuracy of plus or minus 5 mm or 15 per cent is suggested as being realistic in the British Standard. Recent developments in covermeters are now improving on this, with the Protovale CM9, for example, showing better than 8 per cent with an average of better than 2 per cent over a wide range of bar sizes, lapped bars, etc.

The standard also lists a number of extraneous factors which are potential sources of error. Those concerned with magnetic effects from the aggregates or the concrete matrix, and those due to variations in cross-sectional shape of the bars should not affect the modern covermeter, but care must always be taken when dealing with multiple bars (Aldred, 1993) and the effects of adjacent steel such as window frames or scaffolding as mentioned earlier.

6.4.4 Ultrasonic pulse velocity measurement (PUNDIT)

(From the PUNDIT Manual by CNS Electronics.)

Theory of UPV measurement
Introduction
The velocity of ultrasonic pulses travelling in a solid material depends on the density and elastic properties of that material. The quality of some materials is sometimes related to their elastic stiffness so that measurement of ultrasonic pulse velocity in such materials can often be used to indicate their quality as well as to determine their elastic properties. Materials which can be assessed in this way include, in particular, concrete and timber but exclude metals.

When ultrasonic testing is applied to metals its object is to detect internal flaws which send echoes back in the direction of the incident beam and these are picked up by a receiving transducer. The measurement of the time taken for the pulse to travel from a surface to a flaw and back again enables the position of the flaw to be located.

Such a technique cannot be applied to heterogeneous materials like concrete or timber since echoes are generated at the numerous boundaries of the different phases within these materials resulting in a general scattering of pulse energy in all directions.

Velocity of longitudinal pulses in elastic solids
It can be shown that the velocity of a pulse of longitudinal ultrasonic vibrations travelling in an elastic solid is given by the following equation:

$$V = \sqrt{\frac{E}{\rho}} \times \frac{(1-v)}{(1+v)(1-2v)}$$

where E is the dynamic elastic modulus
 ρ is the density
 ν is Poisson's ratio

Effect of size and shape of specimen tested
The above equation may be considered to apply to the transmission of longitudinal pulses through a solid of any shape or size provided the least lateral dimension (i.e. the dimension measured perpendicular to the path travelled by the pulse) is not less than the wavelength of the pulse vibrations. The pulse velocity is not affected by the frequency of the pulse so that the wavelength of the pulse vibrations is inversely proportional to this frequency. Thus the pulse velocity will generally depend only on the properties of the materials and the measurement of this velocity enables an assessment to be made of the condition of the material.

Frequency of pulse vibrations
The pulse frequency used for testing concrete or timber is much lower than that used in metal testing. The higher the frequency, the narrower the beam of pulse propagation but the greater the attenuation (or damping out) of the pulse vibrations.

Metal testing requires high-frequency pulses to provide a narrow beam of energy but such frequencies are unsuitable for use with heterogeneous materials because of the considerable amount of attenuation which pulses undergo when they pass through these materials.

The frequencies suitable for these materials range from about 20 kHz to 250 kHz, with 50 kHz being appropriate for the field testing of concrete. These frequencies correspond to wavelengths ranging from about 200 mm (for the lower frequency) to about 16 mm at the higher frequency.

Method of testing
For assessing the quality of materials from ultrasonic pulse velocity measurement, it is necessary for this measurement to be of a high order of accuracy. This is done using an apparatus which generates suitable pulses and accurately measures the time of their transmission (i.e. transit time) through the material tested. The distance which the pulses travel in the material (i.e. the path length) must also be measured to enable the velocity to be determined from the path lengths and transit times should each be measured to an accuracy better than ± 1 per cent.

The instrument indicates the time taken for the earliest part of the pulse to reach the receiving transducer, measured from the time it leaves the transmitting transducer, when these transducers are placed at suitable points on the surface of the material.

Figure 6.14 shows how the transducers may be arranged on the surface of the specimen tested, the transmission being either direct, indirect or semi-direct. The direct transmission arrangement is the most satisfactory one since the longitudinal pulses leaving the transmitter are propagated mainly in the direction normal to the transducer face. The indirect arrangement is possible because the ultrasonic beam of energy is scattered by discontinuities within the material tested but the strength of the pulse detected in this case is only about 1 per cent or 2 per cent of that detected for the same path length when the direct transmission arrangement is used.

Pulses are not transmitted through large air voids in a material and, if such a void lies

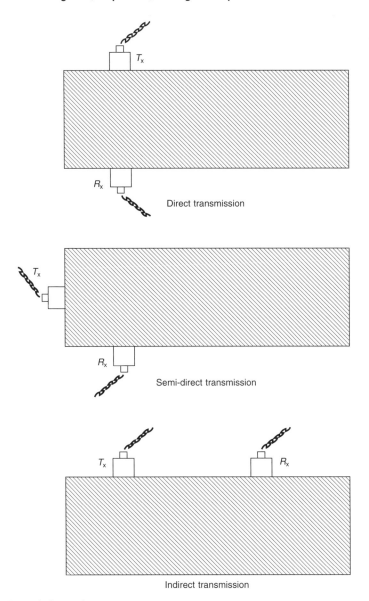

Figure 6.14 Transmission modes.

directly in the pulse path, the instrument will indicate the time taken by the pulse which circumvents the void by the quickest route. It is thus possible to detect large voids when a grid of pulse velocity measurements is made over a region in which these voids are located.

Application of pulse velocity testing

This method of testing was originally developed for use on concrete and the published accounts of its application are concerned predominately with this material.

A considerable volume of literature has been published describing the results of research on the use of ultrasonic testing for concrete and for fuller details of this application, the reader is referred to the References given at the end of this chapter.

In Britain the method was first developed by Jones and Gatfield (1949, 1955) at the Road Research Laboratory between 1945 and 1949 and also independently in Canada by Leslie and Cheesman (1949) at about the same time. The apparatus developed at that time made use of a cathode-ray oscilloscope for the measurement of transit times and modified forms of this equipment have been widely used in many countries. The equipment was particularly useful in the laboratory but was less easy to use under field conditions.

The PUNDIT apparatus used in the UK has been designed particularly for field testing being light, portable and simple to use. It can be operated independently of the mains power supply when used in the field and directly from the AC mains supply for laboratory use.

The British Standards Institution has issued Recommendations for Non-Destructive Methods of Testing for Concrete (BSI, 1974). Part 5 of these, which is concerned with ultrasonic testing, was issued in 1974.

Ultrasonic testing is now widely used in Britain and it is clear that the advantages of this method over traditional methods of testing are likely to increase further its application. In particular its ability to examine the state of concrete in depth is unrivalled.

Testing concrete
Applications

The pulse velocity method of testing may be applied to the testing of plain, reinforced and pre-stressed concrete whether it is pre-cast or cast *in situ*.

Measurement

The measurement of pulse velocity may be used to determine:

(a) the homogeneity of the concrete,
(b) the presence of voids, cracks or other imperfections,
(c) changes in the concrete which may occur with time (i.e. due to the cement hydration) or through the action of fire, frost or chemical attack,
(d) the quality of the concrete in relation to specified standard requirements, which generally refer to its strength.

Accuracy

In most of the applications it is necessary to measure the pulse velocity to a high degree of accuracy since relatively small changes in pulse velocity usually reflect relatively large changes in the condition of the concrete. For this reason it is important that care be taken to obtain the highest possible accuracy of both the transit time and the path length measurements since the pulse velocity measurement depends on both of these.

It is desirable to measure pulse velocity to within an accuracy of ±2 per cent which allows a tolerance in the separate measurements of path length and transit time of only a little more than ±1 per cent.

When such accuracy of path length measurement is difficult or impossible, an estimate of the limits of accuracy of the actual measurements should be recorded with the results so that the reliability of the pulse velocity measurements can be assessed.

Coupling the transducers with the concrete surface

Accuracy of transit time measurement can only be assured if good acoustic coupling between the transducer face and the concrete surface can be achieved. For a concrete surface formed by casting against steel or smooth timber shuttering, good coupling can readily be obtained if the surface is free from dust and grit and covered with a light or medium grease or other suitable couplant. A wet surface presents no problem.

If the surface is moderately rough, a stiffer grease should be used but very rough surfaces require more elaborate preparation. In such cases the surface should be ground flat over an area large enough to accommodate the transducer face or this area may be filled to a level smooth surface with a minimum thickness of a suitable material such as plaster of Paris, cement mortar or epoxy resin, a suitable time being allowed to elapse for the filling material to harden.

If the value of the transit time displayed remains constant to within +1 per cent when the transducers are applied and reapplied to the concrete surface, it is a good indication that satisfactory coupling has been achieved.

Choice of transducer arrangement

Figure 6.14 shows three alternative arrangements for the transducers when testing concrete. Whenever possible, the direct transmission arrangement should be used. This will give maximum sensitivity and provide a well-defined path length. It is, however, sometimes required to examine the concrete by using diagonal paths and semi-direct arrangements are suitable for these.

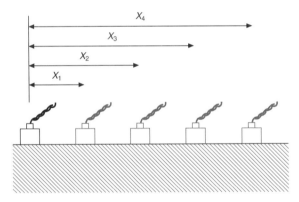

Figure 6.15 Indirect pulse velocity measurement.

The indirect arrangement is the least satisfactory because, apart from its relative insensitivity, it gives pulse velocity measurements which are usually influenced by the concrete layer near the surface and this layer may not be representative of the concrete in deeper layers. Furthermore the length of the path is less well defined and it is not satisfactory to take this as the distance from centre to centre of the transducers. Instead, the method shown in Figure 6.15 should be adopted to determine the effective path length.

In this method, the transmitting transducer is placed on a suitable point on the surface and the receiving transducer is placed on the surface at successive positions along a line and the centre to centre distance is plotted against the transit time. The slope of the straight line drawn through these points gives the mean pulse velocity at the surface.

In general, it will be found that the pulse velocity determined by the indirect method of testing will be lower than that using the direct method. If it is possible to employ both methods of measurement then a relationship may be established between them and a correction factor derived. When it is not possible to use the direct method an approximate value for V_d may be obtained as follows:

$$V_d \text{ approximately equal to } 1.05 V_i$$

where V_d is the pulse velocity obtained using the direct method
       V_i is the pulse velocity obtained using the indirect method

If the points do not lie in a straight line, it is an indication either that the concrete near the surface is of variable quality or that a crack exists in the concrete within the line of the tests position (see below).

A change of slope in the plot could indicate that the pulse velocity near the surface is much lower than it is deeper down in the concrete. This layer of inferior quality could arise as a result of damage by fire, frost, sulfate attack, etc.

In Figure 6.16 for transducer separation distances up to X_0 to the pulse travels through the affected surface layer and the slope of the line gives the pulse velocity in this layer. Beyond X_0 the pulse has travelled along the surface of the underlying sound concrete and the slope of the line beyond X_0 gives the higher velocity in the sound concrete.

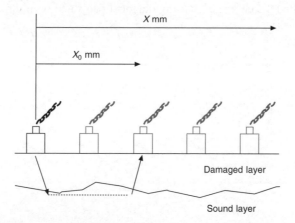

Figure 6.16 Effect of change of material properties.

The thickness of the affected surface layer may be estimated as follows:

$$t = \frac{X_0}{2} \sqrt{\frac{V_s - V_d}{V_s + V_d}}$$

where V_d is the pulse velocity in the damaged concrete
       V_s is the pulse velocity in the underlying sound concrete
       t is the thickness of the layer of damaged concrete
       X_0 is the distance at which the change of slope occurs.

Influence of test conditions

The pulse velocity in concrete may be influenced by:

(a) path length,
(b) lateral dimensions of the specimen tested,
(c) presence of reinforcing steel,
(d) moisture content of the concrete.

The influence of path length will be negligible provided it is not less than 100 mm when 20 mm size aggregate is used or not less than 150 mm for 40 mm size aggregate.

Pulse velocity will not be influenced by the shape of the specimen provided its least lateral dimension (i.e. its dimension measured at right angles **to** the pulse path) is not less than the wavelength of the pulse vibrations. For pulses of 50 kHz frequency, this corresponds to a least lateral dimension of about 80 mm. Otherwise the pulse velocity may be reduced and the results of pulse velocity measurements should be used with caution (Figure 6.17).

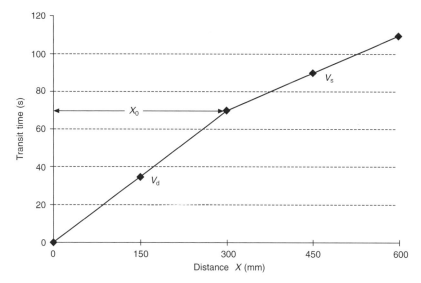

Figure 6.17 Pulse velocity determination by indirect method.

The velocity of pulses in a steel bar is generally higher than they are in concrete. For this reason, pulse velocity measurements made in the vicinity of reinforcing steel may be high and **not** representative of the concrete since the PUNDIT indicates the time for the first pulse to reach the receiving transducer.

The influence of the reinforcement is generally very small if the bars run in a direction at right angles to the pulse path and the quantity of steel is small in relation to the path length.

The PUNDIT manual shows how this influence may be allowed for when the bar diameter lies directly along the pulse path. It is, however, preferable to avoid such a path arrangement and to choose a path which is not in a direct line with the bar diameters.

When the steel bars lie in a direction parallel to the pulse path, the influence of the steel may be more difficult to avoid. Again, it is advisable to choose pulse paths which avoid the influence of the steel as far as possible.

The moisture content of concrete can have a small but significant influence on the pulse velocity. In general, the velocity is increased with increased moisture content, the influence being more marked for lower quality concrete. The pulse velocity of saturated concrete may be up to 2 per cent higher than that in dry concrete of the same composition and quality, although this figure is likely to be lower for high-strength concrete. When pulse velocity measurements are made on concrete as a quality check, a contractor may be encouraged to keep the concrete wet for as long as possible in order to achieve an enhanced value of pulse velocity. This is generally an advantage since it provides an incentive for good curing practice.

The temperature of the concrete has been found to have no significant effect on pulse velocity over the range from 0°C to 30°C so that, except for abnormally extreme temperatures, temperature influence may be disregarded.

Homogeneity of the concrete
Measurement of pulse velocities at points on a regular grid on the surface of a concrete structure provides a reliable method of assessing the homogeneity of the concrete. The size of the grid chosen will depend on the size of the structure and the amount of variability encountered.

It is useful to plot a diagram of pulse velocity contours from the results obtained since this gives a clear picture of the extent of variations. It should be appreciated that the path length can influence the extent of the variations recorded because the pulse velocity measurements correspond to the average quality of the concrete along the line of the pulse path and the size of concrete sample tested at each measurement is directly related to the path length.

Detection of defects
When an ultrasonic pulse travelling through concrete meets a concrete–air interface, there is a negligible transmission of energy across this interface so that any air-filled crack or void lying directly between the transducers will obstruct the direct beam of ultrasound when the void has a projected area larger than the area of the transducer faces. The first pulse to arrive at the receiving transducer will have been diffracted around the periphery of the defect and the transit time will be longer than in similar concrete with no defect.

It is sometimes possible to make use of this effect for locating flaws, etc. but it should be appreciated that small defects often have little or no effect on transmission times.

Detection of large voids or cavities
A large cavity may be detected by measuring the transit times of pulses passing between the transducers when they are placed in suitable positions so that the cavity lies in the direct path between them. The size and position of such cavities may be estimated by assuming that the pulses pass along the shortest path between the transducers and around the cavity. Such estimates are more reliable if the cavity has a well defined boundary surrounded by uniformly dense concrete. If the projected area of the cavity is smaller than the diameter of the transducer, the cavity cannot be detected.

Estimating the depth of surface cracks
An estimate of the depth of a crack visible at the surface can be obtained by measuring the transit times across the crack for two different arrangements of the transducers placed on the surface. One suitable arrangement is one in which the transmitting and receiving transducers are placed on opposite sides of the crack and distant from it. Two values of

x are chosen, one being twice that of the other, and the transit times corresponding to these are measured.

An equation may be derived by assuming that the plane of the crack is perpendicular to the concrete surface and that the concrete in the vicinity of the crack is of reasonably uniform quality.

A check may be made to assess whether the crack is lying in a plane perpendicular to the surface by placing both transducers near to the crack and moving one of them. Details are given in the PUNDIT manual.

It is important that the distance x be measured accurately and that very good coupling is developed between the transducers and the concrete surface. The method is valid provided the crack is not filled with water.

Monitoring changes in concrete with time

Changes occurring in the structure of concrete with time caused by either hydration (which increases strength) or by an aggressive environment, such as frost, or sulfates, may be determined by repeated measurements of pulse velocity at different times. Changes in pulse velocity are indicative of changes in strength and their measurement can be made over progressive periods of time on the same test piece or concrete product.

This facility is particularly useful for following the hardening process during the first two days after casting and it is sometimes possible to take measurements through formwork before it is removed at very early ages. This has a useful application for determining when formwork can be removed or when pre-stressing operations can proceed.

Estimation of strength after fire damage

Pulse velocity measurements may be used to assess the extent of damage to concrete after a fire. Figure 6.18 shows that a good correlation could be obtained to estimate the residual crushing strength of the concrete after heating, from pulse velocity tests.

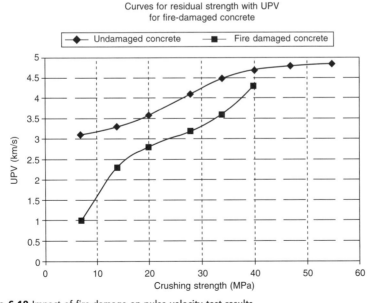

Figure 6.18 Impact of fire damage on pulse velocity test results.

Figure 6.18 further shows that, for a given residual strength, the pulse velocity was apparently less for damaged than that for undamaged concrete. These results were for concrete made with gravel aggregate and are typical of normal concrete although no information is available regarding the effect of different types of aggregate on the correlations.

Estimation of strength

Concrete quality is generally assessed by measuring its cube (or cylinder) crushing strength. It has been found that there is no simple correlation between cube strength and pulse velocity but the correlation is affected by:

(a) type of aggregate,
(b) aggregate/cement ratio
(c) age of concrete size and grading of aggregate
(d) curing conditions

More details of the effects of these can be found in Jones (1949, 1962, 1955 with Gatfield) and Facaoaru (1969).

In practice, if pulse velocity results are to be expressed as equivalent cube strengths, it is preferable to calibrate the particular concrete used by making a series of test specimens with materials and mix proportions the same as the specified concrete but having a range of strengths. The pulse velocity is measured for each specimen which is then tested to failure by crushing.

The range of strength may be obtained either by varying the age of the concrete at test or by introducing a range of water–cement ratios. The curve relating cube strength to pulse velocity is not likely to be the same for these two methods of varying strength but the particular method chosen should be appropriate to the test purpose required.

If strength monitoring with time is to be carried out, the calibration curve is best obtained by varying the age but a check on quality at a particular age would require the correlation to be obtained by varying the water–cement ratio

Although such correlations can be obtained from tests on cubes, it is preferable to use beams such as those used for testing the modulus of rupture of concrete and described in BS 1881 (Methods of testing concrete). These beams are 500 mm long and a more accurate value of pulse velocity is obtained by using the long axis as the pulse path. After testing ultrasonically, the beams are tested in flexure to determine the modulus of rupture and the broken halves tested by crushing to measure the equivalent cube strength. All the details of these tests are described in BS 1881.

Figure 6.19 shows a typical curve obtained for a concrete made with river gravel aggregate and calibrated by using beams.

When testing the concrete in a structure, it would be unreasonable to expect the value of the cube strength estimated from pulse velocity measurements to be the same as that specified for the control cubes made on the site since the design of concrete structures takes into account the fact that cubes are likely to be of higher strength than the concrete in the structure which it represents. A suitable tolerance is therefore required to allow for this. This subject is discussed more fully in Elvery and Din (1969) and Elvery (1971).

Instead of expressing the strength in terms of cube strength, it is preferable to obtain a direct correlation between the strength of a structural member and the pulse velocity whenever this is possible. Such correlation can often be readily applied to pre-cast units and it is possible to obtain a curve relating pulse velocity with an appropriate mechanical test such as bending strength, for the unit.

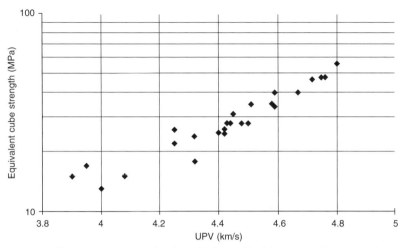

Figure 6.19 Correlation between strength and UPV.

6.4.5 Chemical tests

Chemical analysis of concrete can provide extremely useful information regarding the cause or causes of failure of concrete. The tests most frequently carried out are listed below:

1. Chloride content
2. Cement content
3. Depth of carbonation
4. Sulfate content
5. Type of cement
6. Alkali content

In the following text, an explanation of the reason why each parameter is important is given, followed by an explanation of the test itself.

Chloride content

Chloride when present in reinforced concrete can cause very severe corrosion of the steel reinforcement. Chlorides can originate from two main sources:

(a) 'Internal' chloride, i.e. chloride added to the concrete at the time of mixing. In this category, calcium chloride-accelerating admixtures, contamination of aggregates and the use of sea water or other saline-contaminated water are included.
(b) 'External' chloride, i.e. chloride ingressing into the concrete post-hardening. In this category, both de-icing salt as applied to many highway structures and marine salt, either directly from sea water in structures such as piers, or in the form of air-borne salt spray in structures adjacent to the coast.

The effect of chloride salts depends to some extent on the method of addition. If the chloride is present at the time of mixing, the calcium aluminate (C_3A) phase of the cement will react with the chloride to some extent, chemically binding it as calcium chloroaluminate. In this form, the chloride is insoluble in the pore fluid and is not available to take part in damaging corrosion reactions. The ability of the cement to complex the chloride is limited, however, and depends on the type of cement. Sulfate resisting cement, for example, has a low C_3A content and is therefore less able to complex the chlorides. In any case, experience suggests that if the chloride exceeds about 0.4 per cent by mass of cement, the risk of corrosion increases. This does not automatically mean that concretes with chloride levels higher than this are likely to suffer severe reinforcement corrosion: this depends on the permeability of the concrete and on the depth of carbonation in relation to the cover provided to the steel reinforcement.

When the concrete carbonates, by reaction with atmospheric carbon dioxide, the bound chlorides are released. In effect this provides a higher concentration of soluble chloride immediately in front of the carbonation zone. Normal diffusion processes then cause the chloride to migrate into the concrete. This process, and normal transport of chlorides caused by water soaking into the concrete surface, is responsible for the effect sometimes observed where the chloride level is low at the surface, but increases to a peak a short distance into the concrete (usually just in front of the carbonation zone). The increase in unbound chloride means that more is available to take part in corrosion reactions, so the combined effects of carbonation and chloride are worse than either effect alone.

Passivation of the steel reinforcement in concrete normally occurs due to a two component system comprising a portlandite layer and a thin pH-stabilized iron oxide/hydroxide film on the metal surface (Leek and Poole, 1990). When chloride ions are present, the passivity of the system is lost by dissolution of the portlandite layer, followed by debonding of the passive film. Physical processes operating inside the passive film may also contribute to its disruption.

The critical chloride content required to initiate corrosion depends on whether the chloride was present at the time of mixing, or has ingressed post-hardening, as discussed above. Clearly this also depends on the microclimate of the concrete (temperature and humidity) and also whether the concrete has carbonated. A figure of about 0.2 per cent by mass of cement is generally accepted for chloride ingressing post-hardening and 0.4 per cent (BRE, 1982) or 0.5 per cent (Schiessl and Raupach, 1990) by mass of cement, where chlorides are added at the time of mixing. Good quality concrete can often show a remarkable tolerance for chloride without significant damage, however, at chloride contents up to about 1 per cent by mass of cement (usually for chloride added at the time of mixing: reinforced concrete is much less tolerant of ingressed chloride).

When chlorides have ingressed from an external source particularly in conditions of saturation and low oxygen availability, insidious pitting corrosion can occur, causing massive localized loss of cross-section. This can occur in the early stages without disruption of the concrete underneath.

Test methods

The generally accepted method of test for chloride in hardened concrete is described in BS 1881: Part 124. The test involves crushing a sample of the concrete to a fine dust, extracting the chloride with hot dilute nitric acid and then adding silver nitrate solution to precipitate any chloride present. Ammonium thiocyanate solution is then titrated against the remaining silver and the amount of chloride determined from the difference between the added silver nitrate and that remaining after precipitating the chloride.

Faster and more precise methods based on ion selective electrodes are now available (Grantham, 1993a, b).

Cement content

It is a fundamental requirement of good-quality concrete that it contains an adequate cement content, or more precisely, a sufficiently low water/cement ratio, to provide adequate durability for the intended exposure conditions. In the absence of chemical admixtures, a certain amount of water is required to provide an adequate workability; essentially to simply lubricate the aggregate particles and the cement. To achieve the desired water/cement ratio, the amount of cement required is therefore automatically defined. This can be altered only by changing the physical properties of the aggregate, or by the addition of a water-reducing admixture.

If the cement content is too low (i.e. the water/cement ratio too high) the concrete will be attacked by the weather and be liable to freeze–thaw attack and the effects of carbonation (Neville, 1981). If the cement content is too high, heat of hydration can cause thermal cracking in large pours, the risk of shrinkage increases (because of the higher water content) making curing doubly important, and, if a high-alkali cement is used, the risk of ASR increases with susceptible aggregates (Concrete Society, 1987).

Test methods

The test to determine the cement content of concrete is given in BS 1881:Part 124:1988. It requires the crushed concrete to be extracted with dilute acid and dilute alkali solution to remove the cement. The extract is then analysed for soluble silica and calcium oxide, being the two major components (expressed as oxides) of Portland cement. The cement content is determined by simple proportion from the two parameters. Where soluble components from the aggregate interfere by contributing to the calcium content (e.g. if a limestone aggregate is present) then the silica value would be used for the cement content determination. Conversely, if the silica value was inflated by some soluble component other than the cement, the lime value would be used, provided the analyst was confident that this was unaffected by soluble components from the aggregate. In practice, it is normal to analyse control samples of the aggregate, where these are available, to avoid these problems. With control samples, an accuracy of better than plus or minus 25 kg/m^3 is readily achievable.

Where cement replacement materials such as pfa (pulverized-fuel ash) and ggbfs (ground-granulated blastfurnace slag) are present, the situation is more complex.

Nevertheless, accurate results can often be obtained using total analyses by, for example, X-ray fluorescence methods and applying a simultaneous equations approach (Grantham, 1994).

6.4.6 Depth of carbonation

In a normal, good-quality reinforced concrete, the steel reinforcement is chemically protected from corrosion by the alkaline nature of the concrete. This alkalinity causes the formation of a passive oxide layer around the steel reinforcement. Concrete, however, reacts with atmospheric carbon dioxide (and sulfur dioxide) to cause gradual neutralization of the alkalinity from the surface inwards: a process known as carbonation. The rate at which this occurs is a function of concrete quality, mainly the water/cement ratio and the compaction. It is generally accepted that the rate of the carbonation reaction is inversely proportional to the square root of the age of the structure. If the depth of carbonation is taken in mm and the age of the structure in years, the constant of proportionality is approximately unity.

So for K (Rate constant) = 1

i.e. $$\text{Rate of carbonation (mm/yr)} = \frac{1}{(\text{Age in years})^{0.5}}$$

(*Note:* The rate applies *only* at the particular age chosen. The rate cannot be used for other ages)

Or $$\text{Depth of carbonation (mm)} = (\text{Age in years})^{0.5}$$

Recent research (Parrott, 1994) suggests that the square root relationship holds only at about 50 per cent RH. At higher humidities the power function drops off, so that above 90 per cent RH the depth of carbonation is likely to equate to the (Age in years)$^{0.3}$ and continues to fall at higher humidity. The effect of this is to mean that the carbonation depth will be lower for concrete continuously exposed to higher humidities.

On this basis, even with a cover of only 10 mm, steel reinforcement should be safe for more than 100 years. In practice, however, carbonation often occurs rather faster, either because the concrete is excessively permeable or due to microcracking in the concrete providing secondary paths to the steel other than by normal diffusion processes. Excessive permeability can result from a high water/cement ratio, but can also result from poor curing of the cover concrete. Most modern specifications fail to recognize the importance of curing on concrete quality.

For the reasons given above, the advice given in Table 3.4 of BS 8110, for example, is rather more stringent, in recognition that concrete in practice is often less than perfect.

The depth of carbonation can be measured on a freshly exposed section of the concrete, such as a core, by spraying with an indicator spray such as phenolphthalein. This turns to a pink colour when the concrete is alkaline (above pH 9.2) but remains colourless where the concrete is carbonated, usually as a more or less even zone extending to some depth from the surface. It should be noted that the pH at which the colour of phenolphthalein changes is lower than that at which passivity is lost (which occurs progressively below about pH 11). The test is described in BRE Information Sheet IP 6/81 (BRE, 1981).

It should be noted that carbonation along microcracks and along diffusion paths in poorly compacted concrete, or so called reconstituted stone, may not be readily revealed by the phenolphthalein spray method. Petrographic methods can reveal carbonation of this kind and are recommended.

Sulfate content

Exposure of concretes made with Portland cement to sulfate salts can cause damage due to an expansive reaction between the tricalcium aluminate phase of the cement and the sulfate salt to form crystals of ettringite. Given adequate space to form, the ettringite forms needle like crystals, but in confined space causes an expansive reaction as the amorphous product develops.

True sulfate attack is relatively rare, and research work suggests that concrete made with a reasonable cement content (at least 330 kg/m^3) and a reasonably low water/cement ratio, is attacked only very slowly (but see the section on thaumasile).

The most damaging salts are the more soluble sulfates based on magnesium or sodium sulfates. Calcium sulfate (gypsum) is only sparingly soluble and is less likely to cause damage. The rate of damage is also dependent on the rate of replenishment of the sulfate salts and hence on groundwater movement (Neville, 1981).

Test methods
Sulfate is usually determined by the method given in BS 1881: Part 124: 1988. This involves an acid extraction and precipitation of the sulfate as barium sulfate with barium chloride solution. The resulting barium sulfate is filtered and weighed to determine sulfate gravimetrically.

Methods based upon ion selective electrodes and ion chromatography have also been employed.

High alumina cement
HAC achieved some notoriety during the 1970s following the collapse of several buildings in which it had been used. This was due to a conversion of the cement from one crystalline form into another, weaker, form. At normal temperatures, the hydration of HAC results in the formation of hydrated calcium monoaluminate (CAH10). Smaller amounts of C2AH8 and hydrous alumina are also formed. However, these hydrated calcium aluminates are metastable and can, at higher temperatures and in the presence of moisture, change to give the stable hydrated calcium aluminate C3AH6. This phenomenon is known as 'conversion', and the amount of the change occurring, 'the degree of conversion'.

At normal temperatures, conversion may take many years but at temperatures in excess of 40°C a considerable amount of conversion can occur within a few months.

Conversion results in a loss of strength, increased porosity and reduced resistance to chemical attack. Recently, there has been increasing concern regarding carbonation of high alumina cement concrete. Following conversion, the increased porosity may permit rapid carbonation of the concrete, removing alkaline protection to the steel reinforcement, which may then suffer from corrosion.

Test methods
A test was devised by the Building Research Station to show whether HAC was likely to be present in a concrete (BRE Information Sheet IS15/74). It essentially tests for a significant content of soluble aluminium in solution, following extraction with dilute sodium hydroxide solution.

The presence of the carbonate minerals render any determination of the degree of conversion of the concrete potentially inaccurate. The best procedures for examination of HAC are petrographic and X-ray diffraction analyses.

6.4.7 Compressive strength determination

The compressive strength of a concrete has traditionally been used as a specification tool by engineers to specify the quality of concrete delivered to a job. It has been used for many years because it is a relatively simple property to measure, unlike chemical analysis which requires a well-equipped chemistry laboratory to determine whether an adequate cement content has been used. On-site, a simple slump test will normally be carried out to show whether the concrete complies with the specified workability.

The concrete will normally be specified as a 'designed mix', meaning that the engineer will have asked for a concrete of a certain strength, either to ensure that he has adequate strength for structural purposes, or, more usually, to ensure the overriding requirement that the concrete has adequate durability. The first thing to realize when measuring concrete strength is that concrete is not a uniform material and, when placed, it is even less so. Even with good quality control, a spread of results of 16 N/mm^2 is likely with a Grade 40 concrete. This means that to be sure of getting most (95 per cent) of his results above 40 N/mm^2, the concrete supplier will aim for a target mean strength of some 48 N/mm^2.

These days, BS EN 206-1 allows the specifier to call for an 'equivalent strength class' concrete, meaning that he can specify a strength grade that will give him a reasonable guarantee that he has achieved his desired minimum cement content and/or maximum water/cement ratio. The strength so specified will usually be higher than that which would be called for strength purposes alone. Alternatively, the specifier may call for a 'designated mix', so that the concrete supplied will be suitable for the intended purpose, without the specifier needing to think about cement content, water/cement ratio etc. For example, the specifier wanting a concrete suitable for a domestic driveway would call for a PAV1 mix. This automatically requires a minimum strength of 35 N/mm^2, a minimum cement content of 300 kg/m^3 and a maximum water/cement ratio of 0.6, with air entrainment.

It must be appreciated that the results gained from concrete core samples will not be directly comparable with those from the original cube tests. Cube samples are fully compacted and have been stored under ideal curing conditions for 28 days, prior to test. Core samples, on the other hand, have been taken from *in-situ* concrete, cured in the structure, often inadequately, with perhaps less than perfect compaction. There are also settlement effects, with results on cores from the bottom of a wall or column differing by 15–30 per cent from those of cores taken near the top (BS 6089). Not surprisingly, there are large differences in the strength measured on cores and on cubes made from the original concrete, as supplied. Concrete Society Technical report No. 11 (Concrete Society, 1987) gives excellent guidance to anyone wishing to find their way through the maze of measuring strength in concrete. It provides guidance on planning an investigation, interpreting the results and comparing the test values with those from the original concrete as delivered. Further work has been published to cover concrete made with ggbfs and pfa. An article in *Concrete* magazine covers the subject in summary (Grantham, 1993).

Measuring concrete strength

Core samples are the most common form of sample for this purpose, removed from the structure by diamond drilling. Typically cores will be 100 mm in diameter, and should ideally be at least three times the maximum aggregate size in diameter.

The cores are usually visually described and photographed, concentrating especially on compaction, distribution of aggregate, presence of steel etc. and then trimmed to a length to diameter ratio approaching 1:1. Various methods are available to ensure accuracy of the ends of the core; grinding to a perpendicular flat surface, capping with high alumina cement mortar and capping with hot sulfur are all methods which are used.

The capped core is then crushed (after appropriate curing for HAC capped specimens) in a calibrated compression testing machine. The resulting failure load is converted first to a cylinder strength and second, to an equivalent *in-situ* cube strength. This will generally be some 75 per cent or so of the original cube result, depending on where the core was taken, before allowing for any additional correction due to compaction effects. The test is described in BS EN 12350.

6.4.8 Petrographic examination

Preliminary examination
The samples are examined with the binocular microscope as received and their dimensions and main features are recorded. The features observed include the following:

(a) The presence and position of reinforcement.
(b) The extent to which reinforcement is corroded.
(c) The nature of the external surfaces of the concrete.
(d) The features and distribution of macro and fine cracks.
(e) The distribution and size range and type of the aggregate.
(f) The type and condition of the cement paste.
(g) Any superficial evidence of deleterious processes affecting the concrete.

Polished surfaces
A plate is cut, where possible, from each sample. This is typically about 20 mm thick and usually provides as large a section of the sample as is possible. The plate is polished to give a high quality surface that can be examined with a high quality binocular microscope or even with the petrological microscope if necessary. The polished plate is used to assess the following:

(a) The size, shape and distribution of coarse and fine aggregate.
(b) The coherence, colour, and porosity of the cement paste.
(c) The distribution, size, shape, and content of voids.
(d) The composition of the concrete in terms of the volume proportions of coarse aggregate, fine aggregate, paste and void.
(e) The distribution of fine cracks and microcracks. Often the surface is stained with a penetrative dye, so that these cracks can be seen. Microcrack frequency is measured along lines of traverse across the surface.
(f) The relative abundance of rock types in the coarse aggregate is assessed.

Thin sections
A thin section is prepared for each sample as appropriate. The section is usually made from a plate cut at right angles to the external surface of the concrete, so that the outer 70 mm or so of the concrete is included in the section. Sometimes it is more appropriate to make the section from inner parts of the concrete. This might be appropriate where specific problems are being investigated, for example. The section normally measures about 50×70 mm. Large area thin sections can be useful in some cases.

In manufacturing the thin section, a plate some 10 mm thick is cut from the sample. This is impregnated with a penetrative resin containing a yellow fluorescent dye. The resin penetrates into cracks, microcracks, and capillary pores in the sample. One side of the impregnated plate is then polished and the plate is mounted onto a glass slide. The surplus sample is then removed and the plate is ground and polished to give a final thickness of between 20 and 30 micrometres. At all stages, the cutting and grinding is carried out using an oil-based coolant in order to prevent further hydration of the cement and excessive heating of the section. The thin section is covered and then examined with a high-quality Zeiss petrological photomicroscope. The thin section supplies the following types of information:

(a) Details of the rock types present in the coarse and fine aggregate and in particular structures seen within those rocks.
(b) Details of the aggregate properties are measured such as the degree of strain in quartz.
(c) The size, distribution and abundance of phases in the cement paste are assessed including, for example, the occurrence of calcium hydroxide and the amount of residual unhydrated clinker.
(d) The presence of cement replacement phases such as slag or pfa can usually be recognized (and the amount of these phases can be judged reasonably accurately). The presence of high alumina cement can be detected and the type of cement clinker can often be assessed.
(e) Any products of processes of deterioration of either the cement paste or the aggregate can be recognized.

Broken surfaces

After the specially prepared surfaces and sections are completed, the remainder of the core is examined with the binocular microscope. In particular, the pieces are broken to produce fresh surfaces. These surfaces allow the contents of voids to be studied and the nature of aggregate surfaces or crack surfaces to be investigated.

Composition

The composition of the sample is measured using either the polished slice or the thin section, depending on the size of the sample and on details of the aggregate type and paste. The thin section is preferable, for example where large quantities of dust are present. The volume proportions are found by the method of point counting using a mechanical stage. The amount of coarse aggregate can also be assessed by this method if a distinction can be made between coarse and fine aggregate. The results obtained usually represent the sample reasonably, but may not represent the concrete.

The amount of individual rock types present in the aggregate as a whole are assessed and the saturated density of the sample is measured by the method of immersion in water using vacuum impregnation to ensure saturation. From this information the volume proportions, and the weight fractions of aggregate, cement and water can be calculated.

Water/cement ratio

The hydrating processes of cement paste vary significantly with the original water/cement ratio. Concretes with a low water/cement ratio tend to leave substantial quantities of unhydrated cement clinker and to develop only limited amounts of coarsely crystalline calcium hydroxide. In particular, the extent to which calcium hydroxide is separated into layers on aggregate surfaces and occurs in voids and on void surfaces varies with the original water/cement ratio. The number and proportion of unhydrated cement clinker particles varies inversely with the original water/cement ratio. Comparison with standard concretes made with known water/cement ratios visually, and by measurement allows the water/cement ratio of the cement paste to be assessed directly. The standard error attached to the estimation of water/cement ratio by this means is considered to be approximately ±0.03.

6.4.9 Surface hardness methods

Rebound hammer – Schmidt hammer

One of many factors connected with the quality of concrete is its hardness. Efforts to measure the surface hardness of mass of concrete were first recorded in the 1930s; tests were based on impacting the concrete surface with a specified mass activated by a standard amount of energy. Early methods involved measurements of the size of indentation caused by a steel ball either fixed to a pendulum or spring hammer, or fired from a standardized testing pistol. Later, however, the height of rebound of the mass from the surface was measured. Although it is difficult to justify a theoretical relationship between the measured values from any of these methods and the strength of a concrete, their value lies in the ability to establish empirical relationships between test results and quality of the surface layer. Unfortunately these are subject to many specific restrictions including concrete and member details, as well as equipment reliability and operator technique.

Indentation testing has received attention in Germany and in former states of the USSR as well as the United Kingdom, but has never become very popular. Pin penetration tests have, however, recently received attention in the USA and Japan. The rebound principle, on the other hand, is more widely accepted. The most popular equipment, the Schmidt Rebound Hammer has been in use worldwide for many years. Recommendations for the use of the rebound method are given in BS 1881: Part 202 (45) and ASTM C805 (46).

Rebound test equipment and operation

The Swiss engineer Ernst Schmidt first developed a practicable rebound test hammer in the late 1940s and modern versions are based on this. A spring-controlled hammer mass slides on a plunger within a tubular housing. The plunger retracts against a spring when pressed against the concrete surface and this spring is automatically released when fully tensioned, causing the hammer mass to impact against the concrete through the plunger. When the spring controlled mass rebounds, it takes with it a rider which slides along a scale and is visible through a small window in the side of the casing. The rider can be held in position on the scale by depressing the locking button. The equipment is very simple to use and may be operated either horizontally or vertically either upwards or downwards.

The plunger is pressed strongly and steadily against the concrete at right angles to its surface, until the spring-loaded mass is triggered from its locked position. After the impact, the scale index is read while the hammer is still in the test position. Alternatively, the locking button can be pressed to enable the reading to be retained or results can automatically be recorded by an attached paper recorder. The scale reading is known as the rebound number, and is an arbitrary measure since it depends on the energy stored in the given spring and on the mass used. This version of the equipment is most commonly used, and is most suitable for concrete in the 20–60 N/mm^2 strength range. Electronic digital reading versions of the equipment are available.

Procedure

The equipment is very sensitive to local variations in the concrete, especially to aggregate particles near to the surface and it is therefore necessary to take 12 readings in the area of interest and to average the results obtained. A recommended procedure can be found in BS 1881: Part 202.

The surface to be measured should be smooth clean and dry but if it is required to take measurements on trowelled surfaces, the surface can be smoothed using the carborundum stone provided with the instrument. This stone is NOT a calibration device and must not be used to receive an impact from the hammer as the stone may shatter.

Theory, calibration and interpretation

The test is based on the principle that the rebound of an elastic mass depends on the hardness of the surface upon which it impinges and in this case will provide information about a surface layer of the concrete defined as no more than 30 mm deep. The results give a measure of the relative hardness of this zone, and this cannot be directly related to any other property of the concrete. Energy is lost on impact due to localised crushing of the concrete and internal friction within the body of the concrete, and it is the latter, which is a function of the elastic properties of the concrete constituents, that makes theoretical evaluation of test results extremely difficult. Many factors influence results but must all be considered if rebound number is to be empirically related to strength.

Factors influencing test results

Results are significantly influenced by all the following factors:

(a) Mix characteristics
   (i) Cement type
   (ii) Cement content
   (iii) Coarse aggregate type
(b) Member characteristics
   (i) Mass
   (ii) Compaction
   (iii) Surface type
   (iv) Age, rate of hardening and curing type
   (v) Surface carbonation
   (vi) Moisture condition
   (vii) Stress state and temperature.

Since each of these factors may affect the readings obtained, any attempts to compare or estimate concrete strength will be valid only if they are all standardized for the concrete under test and for the calibration specimens. These influences have different magnitudes. Hammer orientation will also influence measured values although correction factors can be used to allow for this effect.

Mix characteristics

The three mix characteristics listed above are now examined in more detail.

1. *Cement type* Variations in fineness of Portland cement are unlikely to be significant – their influence on strength correlation is less than 10 per cent. Super sulfated cement, however, can be expected to yield strengths 50 per cent lower than suggested by a Portland cement calibration, whereas high alumina cement concrete may be up to 100 per cent, stronger.
2. *Cement content* Changes in cement content do not result in corresponding changes in surface hardness. The combined influence of strength, workability and aggregate/

cement proportions leads to a reduction of hardness relative to strength as the cement content increases. The error in estimated strength, however, is unlikely to exceed 10 per cent from this cause for most mixes.
3 *Coarse aggregate*   The influence of aggregate type and proportions can be considerable, since strength is governed by both paste and aggregate characteristics. The rebound number will be influenced more by the hardened paste. For example, crushed limestone may yield a rebound number significantly lower than for a gravel concrete of similar strength which may typically be equivalent to a strength difference of 6–7 N/mm^2. A particular aggregate type may also yield different rebound number/strength correlations depending on the source and nature of the aggregate.

Lightweight aggregates may be expected to yield results significantly different from those for concrete made with dense aggregates, and considerable variations have also been found between types of lightweight aggregates. Calibrations can, however, be obtained for specific lightweight aggregates, although the amount of natural sand used will affect results.

Member characteristics
The member characteristics listed above are also to be discussed in detail.

1 *Mass*   The effective mass of the concrete specimen or member under test must be sufficiently large to prevent vibration or movement caused by the hammer impact. Any such movement will result in a reduced rebound number. For some structural members the slenderness or mass may be such that this criterion is not fully satisfied, and in such cases absolute strength prediction may be difficult. Strength comparisons between or within individual members must also take account of this factor. The mass of calibration specimens may be effectively increased by clamping them firmly in a heavy testing machine.
2 *Compaction*   Since a smooth, well-compacted surface is required for the test, variations of strength due to internal compaction differences cannot be detected with any reliability. All calibrations must assume full compaction.
3 *Surface type*   Hardness methods are not suitable for open-textured or exposed aggregate surfaces. Trowelled or floated surfaces may be harder than moulded surfaces, and will certainly be more irregular. Although they may be smoothed by grinding, this is laborious and it is best to avoid trowelled surfaces in view of the likely overestimation of strength from hardness readings. The absorption and smoothness of the mould surface will also have a considerable effect. Calibration specimens will normally be cast in steel moulds which are smooth and non-absorbent, but more absorbent shuttering may well produce a harder surface, and hence internal strength may be overestimated. Although moulded surfaces are preferred for on-site testing, care must be taken to ensure that strength calibrations are based on similar surfaces, since considerable errors can result from this cause.
4 *Age, rate of hardening and curing type*   The relationship between hardness and strength has been shown to vary as a function of time and variations in initial rate of hardening, subsequent curing, and exposure conditions will further influence this relationship. Where heat treatment or some other form of accelerated curing has been used, a specific calibration will be necessary. The moisture state may also be influenced by the method of curing. For practical purposes the influence of time may be regarded as

unimportant up to the age of three months, but for older concretes it may be possible to develop regression factors which take account of the concrete's history.

5   *Surface carbonation*   Concrete exposed to the atmosphere will normally form a hard carbonated skin, whose thickness will depend upon the exposure conditions and age. It may exceed 20 mm for old concrete although it is unlikely to be significant at ages of less than three months. The depth of carbonation can easily be determined by the phenolphthalein spray method. Examination of gravel concrete specimen, which had been exposed to an outdoor city centre atmosphere for six months showed a carbonated depth of only 4 mm. This was not sufficient to influence the rebound number strength relationship in comparison with similar specimens stored in a laboratory atmosphere, although for these specimens no measurable skin was detected. In extreme cases, however, it is known that the overestimate of strength from this cause may be up to 50 per cent, and is thus of great importance. When significant carbonation is known to exist the surface layer ceases to be representative of the concrete within an element.

6   *Moisture condition*   The hardness of a concrete surface is lower when wet than when dry, and the rebound strength relationship will be altered accordingly. A wet surface may lead to an underestimate of strength of up to 20 per cent. Field tests and strength calibrations should normally be based on dry surface conditions, but the effect of internal moisture on the strength of control specimens must not be overlooked.

7   *Stress state and temperature*   Both these factors may influence hardness readings, although in normal practical situations this is likely to be small in comparison with the many other variables. Particular attention should, however, be paid to the functioning of the test hammer if it is to be used under extremes of temperature.

Calibration

Clearly, the influences of the variables described above are so great that it is very unlikely that a general calibration curve relating rebound number to strength, as provided by the equipment manufacturers, will be of any practical value. The same applies to the use of computer data processing to give strength predictions based on results from the electronic rebound hammer unless the conversions are based on case-specific data. Strength calibration must be based on the particular mix under investigation, and the mould surface, curing and age of laboratory specimens should correspond as closely as possible to the in-place concrete. It is essential that correct functioning of the rebound hammer is checked regularly using a standard steel anvil of known mass. This is necessary because wear may change the spring and internal friction characteristics of the equipment. Calibrations prepared for one hammer will also not necessarily apply to another. It is probable that very few rebound hammers used for *in-situ* testing are in fact regularly checked against a standard anvil, and the reliability of results may suffer as a consequence.

The importance of specimen mass has been discussed above; it is essential that test specimens are either securely clamped in a heavy testing machine or supported upon an even solid floor. Cubes or cylinders of at least 150 mm should be used, and a minimum restraining load of 15 per cent of the specimen strength has been suggested for cylinders, and BS 1881(45) recommends not less than 7 N/mm^2 for cubes tested with a type N hammer. Typically the relationship between rebound number and restraining load is such that once a sufficient load has been reached the rebound number remains reasonably constant.

It is well established that the crushing strength of a cube tested wet is likely to be about

10 per cent lower than the strength of a corresponding cube tested dry. Since rebound measurements should be taken on a dry surface, it is recommended that wet cured cubes be dried in the laboratory atmosphere for 24 hours before test, and it is therefore to be expected that they will yield higher strengths than if tested wet in the standard manner.

Other near to surface strength tests

These include the Windsor Probe, the BRE Internal Fracture Tester, various break-off devices and the CAPO and Lok tests used extensively in Scandinavia.

6.4.10 Radar profiling

Over the past decade there has been an increasing usage of sub-surface impulse radar to investigate civil engineering problems and, in particular, concrete structures. Electromagnetic waves, typically in the frequency range 500 MHz to 1.5 GHz, will propagate through solids, with the speed and attenuation of the signal influenced by the electrical properties of the solid materials. The dominant physical properties are the electrical permittivity which determines the signal velocity, and the electrical conductivity which determines the signal attenuation. Reflections and refractions of the radar wave will occur at interfaces between different materials and the signal returning to the surface antenna can be interpreted to provide an evaluation of the properties and geometry of sub-surface features.

Radar systems

There are three fundamentally different approaches to using radar to investigate concrete structures.

(a) Frequency modulation – in which the frequency of the transmitted radar signal is continuously swept between pre-defined limits. The return signal is mixed with the currently transmitted signal to give a difference frequency, depending upon the time delay and hence depth of the reflective interface. This system has seen limited use to date on relatively thin walls.
(b) Synthetic pulse radar – in which the frequency of the transmitted radar signal is varied over a series of discontinuous steps. The amplitude and phase of the return signal is analysed and a time domain synthetic pulse is produced. This approach has been used to some extent in the field and also in laboratory transmission line studies to determine the electrical properties of concrete at different radar frequencies.
(c) impulse radar – in which a series of discrete sinusoidal pulses within a specified broad frequency band are transmitted into the concrete typically with a repetition rate of 50 kHz. The transmitted signal is often found to comprise three peaks, with a well-defined nominal centre frequency.

Impulse radar systems have gained the greatest acceptance for field use and most commercially obtainable systems are of this type. The power output of the transmitted radar signal is very low, and no special safety precautions are needed. However, in the UK a Department of Trade and Industry Radiocommunications Agency licence is required to permit use of investigative radar equipment.

Radar equipment

Impulse radar equipment comprises a pulse generator connected to a transmitting antenna. This is commonly of a bow-tie configuration, which is held in contact with the concrete and produces a divergent beam with a degree of spatial polarization. A centre frequency antenna of 1 GHz is often used in the investigation of relatively small concrete elements, up to 500 mm thick, while a 500 MHz antenna may be more appropriate for deeper investigations. However, a lower frequency loses resolution of detail despite the improved penetration.

An alternative to using surface-contact antennae is to use a focused beam horn antenna with an air gap of about 300 mm between the horn and the concrete surface. These systems have been used in the USA and Canada to survey bridge decks from a vehicle moving at speeds of up to 50 km/h, principally to detect corrosion-induced delamination of the concrete slab. Operational details are provided in ASTM D4748.

Structural applications and limitations

In addition to the assessment of concrete bridge decks radar has been used to detect a variety of features buried within concrete ranging from reinforcing bars and voids to murder victims.

The range of principal reported structural applications is summarised in Table 6.2. Interpretation of radar results to identify and evaluate the dimensions of sub-surface features is not always straightforward. The radar picture obtained often does not resemble the form of the embedded features. Circular reflective sections such as metal pipes or reinforcing bars, for example, present a complex hyperbolic pattern due to the diverging nature of the beam. The use of signal processing can simplify the image but interpretation is still complex. Evaluating the depth of a feature of interest necessitates a foreknowledge of the speed at which radar waves will travel through concrete. This is principally determined by the relative permittivity of the concrete, which in turn is determined predominantly by the moisture content.

Table 6.2 Structural applications of radar

Reliability:	
Greatest	Least
Determine major construction features	
Assess element thickness	
Locate reinforcing bars	
Locate moisture	
Locate voids, honeycombing, cracking	
Locate chlorides	
Size reinforcing bars	
Size voids	
	Estimate chloride concentrations
	Locate reinforcement corrosion

Because of the difficulties of interpretation, surveys are normally conducted by specialists who rely on practical experience and have a knowledge of the limitations of the technique in practical situations. For example, features such as voids can be particularly difficult to detect if located very deep or beneath a layer of closely spaced reinforcing steel. The use of neural networks or 'artificial intelligence' has been used to help with interpretation of complex radar traces.

Radar reflects most strongly off metallic objects or from the interface between two materials with widely differing permittivities. An air filled void in dry concrete, which does not differ very strongly in permittivity from the concrete itself can therefore be difficult to detect, especially if the void is small. The same void filled with water, however, would be much more easily detected.

Water strongly attenuates a radar signal and using a 1 GHz antenna, typical practical penetrations of around 500 mm have been achieved for dry concrete, and 300 mm for water-saturated concrete. If the water is contaminated with salt, penetration is likely to be smaller still.

6.4.11 Acoustic emission

Theory
As a material is loaded, localized points may be strained beyond their elastic limit, and crushing or microcracking may occur. The kinetic energy released will propagate small-amplitude elastic stress waves throughout the specimen. These are known as acoustic emissions, although they are generally not in the audible range and may be detected as small displacements by transducers positioned on the surface of the material.

An important feature of many materials is the Kaiser effect, which is the irreversible characteristic of acoustic emission resulting from applied stress. This means that if a material has been stressed to some level, no emission will be detected on subsequent loading until the previously applied stress level has been exceeded. This feature has allowed the method to be applied most usefully to materials testing, but unfortunately the phenomenon does not always apply to plain concrete. Concrete may recover many aspects of its pre-cracking internal structure within a matter of hours due to continued hydration, and energy will again be released during reloading over a similar stress range.

More recent tests on reinforced concrete beams have shown that the Kaiser effect is observed when unloading periods of up to 2 hours have been investigated. However, it is probable that over longer time intervals the autogenic 'healing' of microcracks in concrete will negate the effect.

Equipment
Specialist equipment for this purpose is available in the UK as an integrated system in modular form and lightweight portable models may be used in the field. The results are most conveniently considered as a plot of emission count rate against applied load (Figure 6.20).

Applications and limitations
It has been reported that as the load level on a concrete specimen increases, the emission rate and signal level both increase slowly and consistently until failure approaches, and there is then a rapid increase up to failure. Whilst this allows crack initiation and propagation to be monitored during a period of increasing stress, the method cannot be used for either individual or comparative measurement under static load conditions. It has also been shown that mature concrete provides more acoustic emission on cracking than young

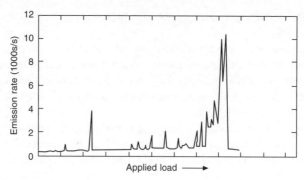

Figure 6.20 Acoustic emission plot.

concrete, but confirms that emissions do not show a significant increase until about 80–90 per cent of ultimate stress. The absence of the Kaiser effect for concrete effectively rules out the method for establishing a history of past stress levels. Hawkins et al. (1979) have, however, described laboratory tests which indicate that it may be possible to detect the degree of bond damage caused by prior loading if the emissions generated by a reinforced specimen under increasing load are filtered to isolate those caused by bond breakdown, since debonding of reinforcement is an irreversible process. Titus et al. (1988) have also suggested that it may be possible to detect the progress of microcracking due to corrosion activity. Long-term creep tests at constant loading by Rossi et al. (1994) have shown a clear link between creep deflection, essentially caused by drying shrinkage microcracking, and acoustic emission levels. However, in tests on plain concrete beams, together with fibre-reinforced and conventional steel-reinforced beams, Jenkins and Steputat (1973) concluded that acoustic emission gave no early warning of incipient failure.

The application to concrete of acoustic emission methods has not yet been fully developed, and as equipment costs are high they must be regarded as essentially laboratory methods at present. However, there is clearly future potential for use of the method in conjunction with *in-situ* load testing as a means of monitoring cracking origin and development and bond breakdown, and to provide a warning of impending failure.

6.4.12 Infrared thermography

This technique uses infrared photographs taken from a structure which has been heated, as it cools. The heating is normally performed by the sun in daytime and the photographs are best recorded in the evening as the structure cools. Infrared thermography offers many potential advantages over other physical methods for the detection of delamination in bridge decks. Areas of sound and unsound concrete will exhibit different thermal characteristics and thus have different surface temperatures as the structure cools. Delaminated areas, for example, will have a different temperature gradient compared to sound areas. Water saturated concrete will appear quite different to dry concrete. The temperature differences are small however, and cannot readily be recorded on infrared film. To view the small temperature differences, a cathode ray tube display is used with the different temperatures recorded as different shades of grey. Thermal contours can, however, be automatically superimposed and colour monitors can be used to give a more graphic picture.

The surface of the structure needs to be viewed from a reasonable distance and so cannot be recorded whilst, for example standing on a deck. Some success has been achieved working from a tall truck at a height of 20 m, provided that the temperature differences were at least 2°C. Working from an aircraft or helicopter, avoids the need for lane closures, but has not consistently shown good results

Holt and Eales (1987) have also described the successful use of thermography to evaluate effects in highway pavements with an infrared scanner and coupled real-time video scanner mounted on a 5 m high mast attached to a van. This is driven at up to 15 mph and images are matched by computer. Procedures for infrared thermography in the investigation of bridge deck delamination are given in ASTM D4788.

Hidden voids or ducts can also sometimes be detected, and techniques have been developed to detect reinforcing bars which have been heated by electrical induction. More recent development of 12-bit equipment has improved the sensitivity to within ±0.1°C. This has enabled high definition imaging and accurate temperature measurement on buildings. The smallest detectable area is reported to be 200×200 mm.

Infrared thermography can also be used to reduce heat losses at hot spots by identifying missing thermal insulation.

6.5 Testing for reinforcement corrosion

6.5.1 Half cell potential testing

Steel embedded in good quality concrete is protected by the high-alkalinity pore water which, in the presence of oxygen, passivates the steel. The loss of alkalinity due to carbonation of the concrete or the penetration of chloride ions (arising from either marine or de-icing salts, or in some cases present *in-situ* from the use of a calcium chloride additive) can destroy the passive film (ACI, 1985; Page and Treadaway, 1982; RILEM 60-CSC; Arup, 1983). In the presence of oxygen and humidity in the concrete, corrosion of the steel starts. A characteristic feature for the corrosion of steel in concrete is the development of macrocells, that is, the co-existence of passive and corroding areas on the same reinforcement bar forming a short-circuited galvanic cell, with the corroding area as the anode and the passive surface as the cathode. The voltage of such a cell can reach as high as 0.5 V or more, especially where chloride ions are present. The resulting current flow (which is directly proportional to the mass lost by the steel) is determined by the electrical resistance of the concrete and the anodic and cathodic reaction resistance (Elsener and Bohni, 1987) (see Figure 6.21).

The current flow in the concrete is accompanied by an electrical field which can be measured at the concrete surface, resulting in equipotential lines that allow the location of the most corroding zones at the most negative values. This is the basis of potential mapping, the principal electrochemical technique applied to the routine inspection of reinforced concrete structures (Stratfull, 1957; Berkeley and Pahmanaban, 1987).

The use of the technique is described in an American Standard, ASTM C876-80, Standard Test Method for Half Cell Potentials of Reinforcing Steel in Concrete.

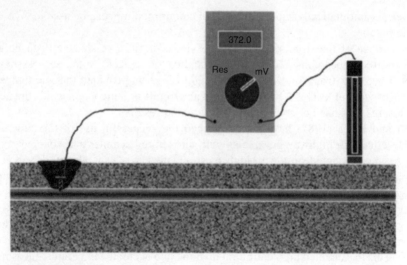

Figure 6.21 Half cell potential.

Factors affecting the potential field

When surface potentials are taken, they are measured remote from the reinforcement due to the concrete cover. The potentials measured are therefore affected by the ohmic drop potential drop in the concrete. Several factors have a significant effect on the potentials measured.

Concrete cover depth

With increasing concrete cover, the potential values at the concrete surface over actively corroding and passive steel become similar. Thus the location of small corroding areas becomes increasingly difficult.

Concrete resistivity

The concrete humidity and the presence of ions in the pore solution affect the electrical resistivity of the concrete. The resistivity may change both across the structure and with time as the local moisture and salt content vary. This may create an error of plus or minus 50 mV in the measured potentials (John *et al.*, 1987).

High resistive surface layers

The macrocell currents tend to avoid highly resistive concrete. The measured potentials at the surface become more positive and corroding areas may be undetected (Berkeley and Pahmanaban, 1987).

Polarization effects

Steel in concrete structures immersed in water or in the earth often have a very negative potential due to restricted oxygen access (Popovics *et al.*, 1983). In the transition region of the structure (splash zone or above ground), negative potentials can be measured due to galvanic coupling with immersed rebars. These negative potentials are not necessarily related to corrosion of the reinforcement.

Procedure for measurement

To measure half cell potentials, an electrical connection is made to the steel reinforcement in part of the member you wish to assess. This is connected to a high-impedance digital millivoltmeter, often backed up with a datalogging device. The other connection to the millivoltmeter is taken to a copper/copper sulfate or silver/silver chloride half cell, which has a porous connection at one end which can be touched to the concrete surface. This will then register the corrosion potential of the steel reinforcement nearest to the point of contact. By measuring results on a regular grid and plotting results as an equipotential contour map, areas of corroding steel may readily be seen. Using 3D mapping techniques, a more graphical representation of the corrosion can be shown, as below.

Results and interpretation

According to the ASTM method, corrosion can only be identified with 95 per cent certainty at potentials more negative than −350 mV. Experience has shown, however, that passive structures tend to show values more positive than −200 mV and often positive potentials. Potentials more negative than −200 mV may be an indicator of the onset of corrosion. The patterns formed by the contours can often be a better guide in these cases. It should be noted that the silver/silver chloride half-cell produces results some 50–70 mV more positive than a copper/copper sulphate cell.

In any case, the technique should never be used in isolation, but should be coupled with measurement of the chloride content of the concrete and its variation with depth and also the cover to the steel and the depth of carbonation (Figures 6.22 and 6.23).

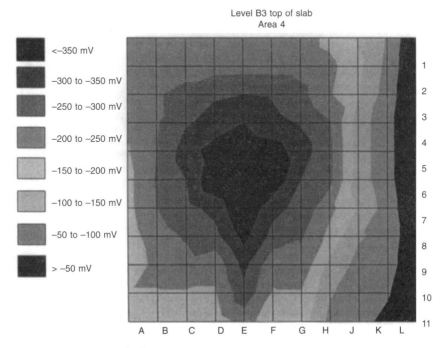

Figure 6.22 Conventional half cell map.

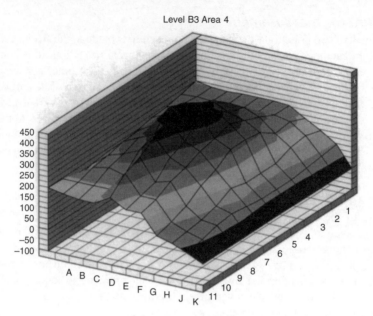

Figure 6.23 3-D half cell map.

6.5.2 Resistivity

The electrical resistivity is an indication of the amount of moisture in the pores, and the size and tortuosity of the pore system. Resistivity is strongly affected by concrete quality, i.e. cement content, water/cement ratio, curing and additives used.

Equipment and use

The main device in use is the four-probe resistivity meter. These have been modified from soils applications and are used by pushing pins directly onto the concrete with moisture or gels to enhance the electrical contact. Millard (Millard *et al.*, 1991) has described two versions of the equipment. Some variations use drilled in probes or a simpler, less accurate two-probe system.

The alternative approach used here measures the resistivity of the cover concrete by a two-electrode method using the reinforcing network as one electrode and a surface probe as the other (Newman, 1966). Concrete resistivity of the area around the sensor is obtained by the formula:

$$\text{Resistivity} = 2\pi \cdot R \cdot D \ (\Omega.\text{cm})$$

where:
- R is the resistance by the 'IR drop' from a pulse between a surface electrode and the rebar network measured by a half cell reference electrode
- D is the electrode diameter of the sensor

Interpretation

Interpretation is empirical. The following interpretations of resistivity measurements have been cited when referring to depassivated steel:

>20 kΩcm   Low corrosion rate
10–20 kΩcm   Low to moderate corrosion rate
5–10 kΩcm   High corrosion rate
< 5 kΩcm   Very high corrosion rate

Researchers working with the GECOR 6 device for resistivity and corrosion rate measurement have conducted laboratory and field research and found the correlation below between resistivity and corrosion rates using the two electrode approach (Broomfield et al., 1994):

>100 kΩcm   Cannot distinguish between active and passive steel
50–100 kΩcm   Low corrosion rate
10–50 kΩcm   Moderate to high corrosion where steel is active
<10 kΩcm   Resistivity is not the controlling parameter

In the above method and interpretation resistivity measurement is used alongside linear polarization measurements (see below), not as a stand-alone technique.

Limitations

The resistivity measurement is a useful additional measurement to aid in identifying problem areas or confirming concerns about poor-quality concrete. Readings can only be considered alongside other measurements.

There is a frequent temptation to multiply the resistivity by the half cell potential and present this as the corrosion rate. This is incorrect. The corrosion rate is usually controlled by the interfacial resistance between the steel and the concrete, not the bulk concrete resistivity. The potential measured by a half cell is not the potential at the steel surface that drives the corrosion cell. Any correlation is fortuitous as described above. However, a high-resistivity concrete will not sustain a high corrosion rate while a low resistivity concrete can, *if* the steel is depassivated and there is sufficient oxygen and moisture present, and if the steel has been depassivated by the presence of chlorides or carbonation.

6.5.3 Corrosion rate

The corrosion rate is probably the nearest the engineer can get to measuring the rate of deterioration with current technology. There are various ways of measuring the rate of corrosion, including AC Impedance and electrochemical noise (Dawson, 1983). However, these techniques are not fieldworthy for the corrosion of steel in concrete so this section will concentrate on linear polarization, also known as polarization resistance as used in the GECOR device.

Property to be measured

It is possible, with varying degrees of accuracy, to measure the amount of steel dissolving and forming oxide (rust). This is done directly as a measurement of the electric current generated by the anodic reaction:

$$Fe \rightarrow Fe^{2+} + 2e^-$$

and consumed by the cathodic reaction:

$$H_2O + {}^1/_2O_2 + 2e^- \rightarrow 2OH^-$$

and then converting the current flow by Faraday's law to metal loss:

$$m = \frac{MIt}{zF}$$

where m = mass of steel consumed
  I = current (amperes, A)
  t = time (seconds, s)
  F = 96 500 A.s
  z = ionic charge (2 for Fe $\rightarrow$ Fe^{2+} + 2e$^-$)
  M = Atomic mass of metal (56 g for Fe)

This gives a conversion of 1 A.cm^{-2} = 11.6 µm per year.

Equipment and use – linear polarization

A typical set-up is shown in Figure 6.24. The system has a rebar connection, a half cell, an auxiliary electrode to apply the perturbing current and a battery operated unit to supply the DC electric field and measure its effect via the half cell. The simplest devices of this type are configured as in Figure 6.25.

Figure 6.24 Linear polarization equipment.

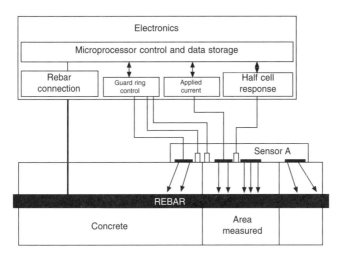

Figure 6.25 Schematic of corrosion rate device.

The potential is measured from the central half cell. Current I is passed from the surrounding electrode to the steel and the potential shift E is measured. This can be repeated for increasing increments of I. The potential must be stable throughout the reading so that a true E is recorded.

The equation for the corrosion current is given in a paper by Stern and Geary (1957):

$$I_{corr} = \frac{B}{R_p}$$

where

I_{corr} is the corrosion current
B is a constant related to the anodic and cathodic Tafel slopes

$$R_p \text{ the polarization resistance} = \frac{\delta E}{\delta I}$$

where I is the change in current
E is the change in potential

The problem with the simple devices is that they can be slow to operate but more importantly they do not define the area of measurement accurately. At low corrosion rates this can lead to errors by orders of magnitude (Fliz et al., 1992).

The GECOR device is more accurate. This works on the linear polarization principle and uses a guard ring. The device has been described in several papers (e.g. Feliú et al.) and the developers have worked on the interpretation of results from carbonated as well as chloride-induced corrosion. Its important features are the two extra half cells that are used to control the guard ring current and define the area of measurement.

In an assessment of three different devices, one without a guard ring, one with a simple guard ring and the device with the sophisticated half cell controlled guard ring, the researchers found good correlation between this device and the most sophisticated laboratory measurements except where the concrete resistivity was very high or cover to the rebar was very deep. Independent field trials also showed good performance.

Gowers *et al.* have used the linear polarization technique with embedded probes (half cell and a simple counter electrode) to monitor the corrosion of marine concrete structures. This technique was described previously without reference to isolating the section of bar to be measured (Langford and Broomfield, 1987). By repeating the measurement in the same location on an isolated section of steel of known surface area the corrosion rate of the actual rebar can be inferred. The main problem is the long term durability of electrical connections in marine conditions.

Interpretation – linear polarization

The following broad criteria for corrosion have been developed from field and laboratory investigations with the sensor-controlled guard ring device:

$I_{corr} < 0.1$ A/cm^2	Passive condition
I_{corr} 0.1 to 0.5 A/cm^2	Low to moderate corrosion
I_{corr} 0.5 to 1 A/cm^2	Moderate to high corrosion
$I_{corr} > 1$ A/cm^2	High corrosion rate

These measurements are affected by temperature and relative humidity (RH), so the conditions of measurement will affect the interpretation of the limits defined above. The measurements should be considered accurate to within a factor of two.

Work has been done in translating I_{corr} to section loss and end of service life (Andrade *et al.*, 1990). However, the loss of concrete is the most usual cause for concern, rather than loss of reinforcement strength. It is far more difficult to predict cracking and spalling rates, especially from an instantaneous measurement. A simple extrapolation assuming that the instantaneous corrosion rate on a certain day is the average rate throughout the life of the structure often gives inaccurate results. This is true for section loss. Converting that to delamination rates is even less accurate as it requires further assumptions about oxide volume and stresses required for cracking the concrete.

The conversion of I_{corr} measurements to damage rates is still being worked on. Another area of concern is when corrosion is due to pitting. Much of the research on linear polarization has been done on highway bridge decks in the USA where chloride levels are high and pitting is not observed. However, in Europe, where corrosion is localized in run down areas on bridge substructures and pitting is more common, the problems of interpretation are complicated as the I_{corr} reading is coming from isolated pits rather than uniformly from the area of measurement. Work has been carried out to assess the damage that would be caused by both generalized and pitting corrosion, assuming a worst case scenario for pitting (Rodriguez *et al.*, 1996).

Laboratory tests with the guard ring device have shown that the corrosion rate can be up to ten times higher than generalized corrosion (Gonzalez *et al.*, 1993). This means that the device is very sensitive to pits. However, it cannot differentiate between pitting and generalized corrosion.

Figure 6.26 illustrates the possible effect of each type of corrosion on 6 mm and 20 mm bars.

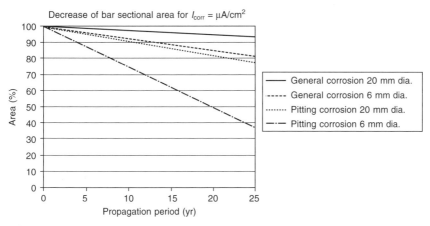

Figure 6.26 Impact of corrosion rates on different bars.

References

ACI Committee 222 (1985) Corrosion of Metals in Concrete. ACIR-85 American Concrete Institute, Detroit, MI.

Aldred, J.A. (1993) Quantifying the losses in cover-meter accuracy due to congestion of reinforcement. *Proceedings of the 5th International Conference on Structural Faults and Repair*, Engineering Technics Press.

Andrade, C., Alonso, M.C. and Gonzalez, J.A. (1990) An initial effort to use corrosion rate measurements for estimating rebar durability. In Berke, N.S. et al. (eds), *Corrosion Rates of Steel in Concrete*, ASTM STP 1065, ASTM, Philadelphia, 29–37.

Arup, H. (1983) The Mechanism of Protection of Steel by Concrete. In Crane, A.P. (ed.), *Corrosion of Reinforcement in Concrete Construction*, Society of Chemical Industry, 151–7.

BS 4408: Part 5: 1974 Recommendations for Non-destructive Methods of Test for Concrete. Part 5. Measurement of the velocity of ultrasonic pulses in concrete. British Standards Institution, London.

Berkeley, K.G. and Pahmanaban, S. (1987) Practical potential monitoring in concrete. *Proc. Conf. UK Corrosion*, (1982) 115–131.

BRE Digest 264 (1982) The Durability of Steel in Concrete: Part 2, Diagnosis and Assessment of Corrosion Cracked Concrete. HMSO, London, August.

BRE (1981) Information Sheet IP 6/81, Carbonation of Concrete Made with Dense Natural Aggregates. HMSO, London.

BRE Information Sheet IS15/74 A Rapid Chemical Test for High Alumina Cement.

Broomfield, J.P., Rodríguez, J., Ortega, L.M. and García, A.M. (1994) Corrosion rate measurements in reinforced concrete structures by a linear polarisation device. Philip D. Cady Symposium on Corrosion of Steel in Concrete American Concrete Institute, Special Publication to be published.

BS 1881: Part 124: Methods for the Chemical Analysis of Hardened Concrete. British Standards Institution, London.

BS 1881: 204: 1988 Recommendations on the Use of Electromagnetic Cover Measuring Devices. British Standards Institution, London.

BS 5328: Method for Specifying Concrete. British Standards Institution, London.

BS 6089: Guide to the Assessment of Concrete Strength in Existing Structures. British Standards Institution, London.

Cement & Concrete Association (1977) The Ultrasonic-Pulse-Velocity Method of Test for Concrete in Structures. Advisory Data Sheet No. 34. Oct.

Chefdeville, J. and Dawance, G. (1950) L'auscultation dynamique du beton. *Annales de l'Institut Technique du Bâtiment et des Travaux Publics.* No. 140, July–Aug.

Chung, H.W. (1978) An appraisal of the ultrasonic pulse technique for detecting voids in concrete. *Concrete,* **12**, No. 11.

Concrete Society (1987a) Technical report No. 11, Concrete Core Testing for Strength. The Concrete Society.

Concrete Society (1987b) Technical Report No. 30. Alkali Silica Reaction, Minimising the Risk of Damage to Concrete, Guidance Notes and Model Specification Clauses. Concrete Society.

Davis, S.G. and Martin, S.J. (1973) The Quality of Concrete and its Variation in Structures. Cement and Concrete Association, Technical Report 42.487, November.

Davis, W.R. and Brough, R. (1972) Ultrasonic techniques in ceramic research and testing. *Ultrasonics,* May.

Dawson, J.L. (1983) Corrosion monitoring of steel in concrete. In Crane, A.P. (ed.), *Corrosion of Reinforcement in Concrete Construction.* Ellis Horwood, for Soc. of Chem. Industry, 175–192.

Drysdale, R.G. (1973) Variations of concrete strength in existing buildings. *Concrete Research,* **25**, No. 85, December.

Elsener, B. and Bohni, H. (1987) *Scheiwz, Ingenieur und Architekt,* **105**, 528.

Elvery, R.H. (1973) Estimating strength of concrete in structures. Current Practice Sheet No. 10, *Concrete,* **7**, No. 11, November (Journal of the Concrete Society).

Elvery, R.H. and Nwokoye, D.N. (1969) Strength assessment of timber for glued laminated beams. Paper 11, Symposium on Non-destructive Testing of Concrete and Timber, Inst. of Civil Engineers, June, London, 105–113.

Elvery, R.H. (1971) Non-destructive testing of concrete and its relationship to specifications. *Concrete,* **5**, No. 4, May (Journal of the Concrete Society.)

Elvery, R.H. and Din, N. (1969) Ultrasonic inspection of reinforced concrete flexural members. Paper 5. Symposium on Non-destructive Testing of Concrete and Timber, June, Institution of Civil Engineers, London, 51–58.

Elvery, R.H. and Forrester, J.A. (1971) Non-destructive testing of concrete. In *Progress in Construction Science and Technology,* Medical and Technical Publishing Co., Aylesbury.

Facaoaru, I. (1969) Non-destructive testing of concrete in Romania. I Paper 4, Symposium on Non-destructive testing of Concrete and Timber, Institution of Civil Engineers, June, London.

Feliú, S., González, J.A., Andrade, C. and Feliú, V. On-site determination of the polarisation resistance in a reinforced concrete Beam. *Corrosion,* **44**, (10), 761–765.

Fliz, J., Sehgal, D.L., Kho Y-T., Sabotl, S., Pickering, H., Osseo-Assare, K. and Cady, P. D. (1992) *Condition Evaluation of Concrete Bridges Relative to Reinforcement Corrosion. Vol. 2: Method for Measuring the Corrosion Rate of Reinforcing Steel,* National Research Council. Washington DC. SHRP-S-324.

Gonzáles, J.A., Andrade, C., Rodríguez, P., Alonso, C. and Feliú, S. (1993) Effects of corrosion on the degradation of reinforced concrete structures. *Progress in Understanding and Prevention of Corrosion,* Inst. of Materials for European Federation of Corrosion, 629–633.

Gowers, K.R., Millard, S.G. and Gill, J.S. Techniques for increasing the accuracy of linear polarisation measurement in concrete structures.

Grantham, M.G. (1993a) Concrete cube failures, counting the cost. *Concrete,* Sept/Oct, 38–40.

Grantham, M.G. (1993b) An automated method for the analysis of chloride in hardened concrete. *Proceedings of the 1993 Conference on Structural Faults and Repair.* Edinburgh, June. ECS Publications.

Grantham, M.G. (1994) Determination of slag and PFA in hardened concrete – the method of last resort revisited. In Kosmatka, S.H. and Jeknavorian, A.A. (eds), *Determination of the Chemical and Mineral Admixture Content of Hardened Concrete, ASTM STP 1253,* American Society for Testing and Materials, Philadelphia.

Hawkins, N.M., Kobayashi, A.S. and Forney, M.E. (1979) Use of holographic and acoustic emission techniques to detect structural damage in concrete members. *Experimental Methods in Concrete Structures for Practitioners,* ACI, Detroit.

Holt, F.B. and Eales, J.W. (1987) Non-destructive evaluation of pavements. *Concrete International.* **9**, No. 6 June, 41–45.

Jenkins, D.R. and Steputat. (1973) Acoustic emission monitoring of damage initiation and development in structurally reinforced concrete beams. *Proc. Structural Faults and Repair 93*, Eng. Technics Press, Edinburgh, **3**, 79–87.

John, D.G., Eden, D.A., Dawson J.L. and Langford P.E. (1987) *Proc. Conf. Corrosion/87*, San Francisco, (1969) CA, 9.-13.3 Paper 136.

Jones, R. (1949) The non-destructive testing of concrete, *Magazine of Concrete Research*, No. 2. June, 67–78.

Jones, R. (1962) *Non-destructive Testing of Concrete*, Cambridge University Press, Cambridge.

Jones, R. (1969) A review of the non-destructive testing of concrete. Paper 1, Symposium on Non-destructive Testing of Concrete and Timber, Institution of Civil Engineers, June, London, 1–7.

Jones, R. and Gatfield, E.N. (1955) Testing concrete by an ultrasonic pulse technique. Road Research Laboratory, Technical Paper No. 34, HMSO, London.

Kemi, H. and Kurabayashi, S. (1978). Distribution of strength of relatively massive concrete structures. Hokkaido General Meeting of Japan Architectural Institute. Sept.

Langford, P. and Broomfield, J. (1987) Monitoring the corrosion of reinforcing steel. *Construction Repair*, **1**, No. 2. May, 32–36.

Lee, I.D.G. Non-destructive testing of timber helicopter rotor blades. Timber Research and Development Association. Test Record E/TR/15.

Lee, I.D.G. (1969) Testing for safety in timber structures. Paper 12, Symposium on Non-destructive Testing of Concrete and Timber, Inst. of Civil Engineers, London, June, 115–118.

Leek, D.S. and Poole, A.B. (1990) The breakdown of the passive film on high yield mild steel by chloride ions. *Proceedings of the 1990 SCI Conference on Corrosion of Reinforcement in Concrete*. Elsevier, New York.

Leslie, J.R. and Cheesman, W.J. (1949) An ultrasonic method of studying deterioration and cracking in concrete structures. *Proc. American Concrete Institute*, Vol. 46, 17–36.

Miles, C.A. and Cutting, C.L. (1974) Changes in the velocity of ultrasound in meat during freezing. *Journal of Food Technology*, **9**, 119–222.

Millard, S.G., Harrison, J. A. and Gowers, K.R. (1991) Practical measurement of concrete resistivity. *British Journal of NDT*, **33**, No. 2, 59–63, February.

Neville, A.M. (1981) *Properties of Concrete*, 3rd edn, Longman, London.

Newman, J. (1966) *Journal of the Electrochemical Society*, **113**, 501.

Page, C.L. and Treadaway, K.W.J. (1982) *Nature*, **297**, 109.

Parrott, L.J. (1994) Carbonation induced corrosion. In proceedings of a one day seminar *Improving Civil Engineering Structures – Old and New*, Geological Society, London, 31.1.95, Geotechnical Publishing Ltd..

Popovics, S., Simeonov, Y., Bozhinov, G. and Barovsky, N. (1983) In Crane A.P. (ed.), *Corrosion of Reinforcement in Concrete Construction*, Society of Chemical Industry, London, 193–222.

RILEM Technical Committee 60-CSC, State of the Art Report, Corrosion of Steel in Concrete.

Rodriguez, J., Ortega, L.M. *et al.* (1996) Corrosion of reinforcement and service life of concrete structures. *Proceedings 7th Int. Conference on Durability of Building Materials and Components*, Stockholm.

Rossi, P. *et al.* (1994) Investigations of the basic creep of concrete by acoustic emission. *Materials and Structures*, **27**, 510–514.

Samarral, M.A. and Elvery, R.H. (1974) The influence of fibres upon crack development in reinforced concrete subject to uniaxial tension. *Concrete Research*, **26**, No. 89, December.

Schiessl, P. and Raupach, M. (1990) Influence of concrete composition and microclimate on the critical chloride content in concrete. *Proceedings of the 1990 SCI conference on Corrosion of Reinforced in Concrete*. Elsevier, New York.

Stern, M. and Geary, A.L. (1957) Electrochemical polarisation. I. A theoretical analysis of the shape of polarisation curves. *J. Electrochem. Soc*, **104**, 56–63.

Stratfull, R.F. (1957) *Corrosion NACE*, **13**, 173t.
Titus, R.N.K *et al.* (1988) Acoustic emission crack monitoring and prediction of remaining life of corroding reinforced concrete beams. *Proc. 4th European Conference on NDT*, Vol. 2, Pergamon, Oxford, 1031–1040.
Tomsett, H.N. (1979) *The In Situ Evaluation of Concrete using Pulse Velocity Differences*, C & CA, July.
Uemura, Sekine, Wakamatsu, Mogami, Baba, Nakamura and Uenoen (1978) Load-bearing test of heat-damaged reinforced concrete beam, Architectural Research Institute, Ministry of Construction. Hokkaido General Meeting of Japan Architectural Institute. Sept.
Watkeys, D.G. (1967) *Non-Destructive Testing of Concrete Subject to Fire Attack*. Thesis submitted for MSc degree. University College, London.
Whitehurst, E.A. (1951) Soniscope tests concrete structures. *Proc. American Concrete Institute*, Vol. 47, 433.

PART 2
Repair

7

Concrete repairs

Michael Grantham

7.1 Patch repairs

Before approaching concrete repairs, consideration must first be given to the cause of the problem. This is fundamental to the success or failure of the repair, and a lack of adequate attention at this point can jeopardize the whole job.

If the problem has been diagnosed as being due to carbonation-induced or chloride-induced corrosion of the reinforcement, patch repairs may be used, although with the latter, there are precautions needed to ensure a successful repair. This chapter deals with cracking and spalling due to reinforcement corrosion only. If the problem has been diagnosed as due to ASR or any of the other mechanisms of failure, the repair may need to be specifically designed for that particular contract and structure, and it is impossible to generalize about the approach which might be taken.

7.1.1 Patch repairing carbonation-induced corrosion

It must first be understood that carbonation is a variable process and the carbonation front will probably not be uniform over the structure. Regular testing to determine the cover to the steel reinforcement and the penetration of carbonation, by the phenolphthalein spray method, will be required.

Criteria employed in the past have been to establish which areas of concrete fall into the following categories:

(a) Visibly spalled concrete with exposed steel reinforcement.
(b) Areas which sound hollow when tapped lightly with a club hammer (these may often

be found around actual spalled areas showing the problem is worse than is visually apparent).
(c) Areas where the cover to the steel reinforcement is less than 10 mm.
(d) Areas where the carbonation front has encroached to within 5 mm of the reinforcement (this necessitates carbonation testing at least every 2 metres or so).
(e) Areas of honeycombed concrete.
(f) Areas where the half cell potential values numerically exceed –200 mV (copper/copper sulfate). Using this criterion, however, caution must be exercised as half cell potentials only show areas of active corrosion. In the summer, whole areas of corrosion may shut down as the concrete dries out. Detection of carbonation-induced corrosion is not as reliable as for chloride-induced corrosion using the half cell method. If the above criteria have been followed, half cell potential testing is not strictly necessary but can be a useful additional aid to diagnosis.

The concrete surface will usually be water jetted prior to cutting out or lightly grit blasted to remove surface grime and deposits and this will also often reveal blowholes in the concrete surface.

All of the areas identified above will require breaking out behind the steel reinforcement, sufficient to get one's fingers behind the steel, ensuring that the edge of the breakout is cut square and not 'feather edged' which inevitably causes failure of the repairs.

At all locations, the steel must be at very least wire brushed to remove all loose deposits or preferably grit blasted or water jetted back to bright metal. If chlorides are present this is essential and will be dealt with in the next section. If the loss of section of the steel exceeds a critical amount, it may be necessary to splice in additional or replacement reinforcement. This decision will be taken by the engineer.

Once the area to be repaired has been cut out and cleaned thoroughly of all debris and dust, the repair can begin. Usually, purpose-designed repair mortars will be used supplied by one of the leading manufacturers. These usually contain one or more of acrylic or other polymers, shrinkage-compensating additives, silica fume, ggbfs or a variety of other chemical modifiers to improve workability, permeability, or ease of application. Simple sand/cement mortars alone should *never* be used, although gauging sand and cement with, for example, an SBR polymer emulsion has been used and can work provided the mortar is correctly batched and gauged with SBR according to manufacturer's instructions.

It is likely that, for pre-batched proprietary repair mortars, some kind of bonding agent will be specified. These must be applied strictly in accordance with the manufacturer's instructions, which will often call for the repair to be applied while the bonding agent is still wet. If used incorrectly, bonding agents can make very efficient *debonding* agents!

If no bonding agent is used, the broken-out surface of the concrete must be thoroughly dampened to kill the suction, *ensuring that the surface is saturated but surface dry at the time of application of the repair.* The author has been involved in the investigation of many failures where the substrate was not adequately wetted and water has been sucked from the interface of the repair and backing concrete, causing poor hydration of the cement and failure of the repair.

The repair mortar is then mixed up *again following fully any special instructions from the manufacturer.* Repairs are then usually hand placed packing the mortar on in layers no thicker than those recommended and using the fingers to pack mortar behind the bar. Wearing of suitable protective clothing is essential as cementitious repair mortars are very alkaline and can cause burns or dermatitis. Similarly, dust masks must be worn when mixing the mortar. The Health and Safety at Work Act and the more recent CDM regulations in the UK place very specific responsibilities on both site operatives and their supervisors and employers to work safely on-site. Once the repair has been filled, a wooden rule is used to level the repair, and the repair is then floated to a smooth finish with a steel float. The repair *must* then be cured in accordance with the manufacturer's recommendations. This may involve the application of damp hessian or more usually a spray-applied curing membrane.

Once finished and cured, there may be surface irregularities and blowholes and there will almost certainly be colour variation between the patch and the surrounding parent concrete. It is usual therefore to apply a thin skim coat of a proprietary material, usually known as a 'fairing coat' to fill in any surface blemishes and to mask the patches themselves. This is followed by the application of a final coating of anti-carbonation paint which has the dual purpose of stopping further carbonation in unrepaired areas and providing a pleasing and even colour finish. The coating will require periodic renewal at perhaps 10–15-year intervals.

7.1.2 Patch repairing chloride-induced corrosion

The problem with chloride-induced corrosion is that the corrosion mechanism is often not fully understood by engineers and clients. Whereas carbonation-induced corrosion causes relatively large areas of the reinforcement to corrode, chloride contamination usually causes very localized areas of the steel to corrode, which is known as pitting corrosion. This can result in deep 'pits' in the steel surface and considerable localized loss of section. Often the corrosion deposits are black and hard and there appears to be little loss of section at first, until the deposits are either dug out with a penknife, or water jetted clear, when the full extent of damage can be seen.

Repairing chloride-contaminated concrete is similar to that previously written for carbonation-induced corrosion except for several very important points:

(a) All corrosion deposits must be removed by grit blasting or water jetting to bright steel. Failure to do so leaves chloride contamination behind in the corrosion deposits and failure will ensue.
(b) *All* of the chloride contaminated concrete must be removed, not just the areas which have spalled or delaminated. (See comments below on the 'incipient anode effect'.)
(c) A criterion often employed is to repair all areas where chloride exceeds, say, 0.4 per cent or 0.5 per cent of cement. Strictly, for chloride added at the time of mixing, the criterion should be 0.4 per cent or for chloride ingressing post-hardening, 0.2 per cent. For pre-stressed or post-tensioned concrete, the maximum tolerable chloride level is only 0.1 per cent by mass of cement.

7.1.3 The incipient anode effect

When an area of steel is corroding under the influence of chloride contamination, steel is dissolving causing the formation of iron 'ions', tiny charged particles of iron. Simultaneously, electrons are released which flow along the bar and react at some point remote from the corrosion with both air and oxygen. Cathodic protection, which will be discussed in the next section, involves providing a small electric current to the steel which prevents further corrosion. The corroding areas are supplying electrons to surrounding areas of steel effectively providing localized cathodic protection to the adjacent steel.

If the corroding area is now broken out and a patch repair applied, without dealing with chloride contamination in adjacent areas, **the cathodic protection system has been removed**!

New corrosion cells will rapidly spring up on either side of the repair and early failure will often ensue. One manufacturer uses small sacrificial anodes, tied to the steel to minimize this problem, but this will only work while zinc remains, which must be gradually being sacrificed to protect the surrounding steel. It also relies on the field of activity being sufficiently wide to inhibit the adjacent cells.

This of course brings economics into play as it may be uneconomical if the areas are large to cut out all of the contamination or to use large numbers of sacrificial anodes. In this case, desalination or cathodic protection may be viable alternatives.

These repair recommendations are only a guide and apply only to conventionally reinforced structures. Repair of pre-stressed or post-tensioned structures with tendon damage is a specialist task and beyond the scope of this book.

More detailed recommendations, together with specimen bills of quantities, can be found in Concrete Society Technical Report 38 (Concrete Society, 1991).

7.2 Cathodic protection

7.2.1 Basic electrochemistry

A galvanic cell consists of two different metals (electrodes) connected through a conducting solution (an electrolyte) and also connected externally completing a circuit. In such a situation, one of the metals (the more reactive) will tend to dissolve in the electrolyte while the other will tend to have new metal deposited on it. In the process of dissolving, the more reactive metal will liberate electrons which flow via the external connection (as an electric current) to be used in metal deposition at the other electrode. The dissolution and deposition reactions are called half cell reactions; the former is the anode reaction and the latter is the cathode reaction.

7.2.2 Corrosion

Corrosion is a similar process but with electrode sites at adjacent locations on the same metal surface. For the corrosion of steel in concrete, the two half cell reactions are as follows:

1. Iron dissolves forming positively charged iron 'ions' and releasing electrons (e^-)

$$\text{Anodic reaction: } Fe \rightarrow Fe^{2+} + 2e^- \tag{7.1}$$

2. The released electrons migrate along the steel and react with oxygen and moisture to form hydroxyl ions (OH^-)

$$\text{Cathodic reaction: } H_2O + 1/2 O_2 + 2e^- \rightarrow 2OH^- \qquad (7.2)$$

The rate of the corrosion process depends upon the availability of the reactants which are iron (Fe), water (H_2O), oxygen (O_2) and electrons (e^-). Thus, for example, if the system is completely dry or starved of oxygen, the steel may still be active but the corrosion rate will be negligible. The relative availability of electrons is equivalent to the corrosion current which depends upon the electrical resistance of the system.

7.2.3 Reactivity

Whether or not the corrosion process can occur at all depends on the pH of the solution adjacent to the steel which, together with other conditions such as the availability of reactants, determines the electrical potential of the steel. Generally, a metal exists in one of three possible reaction states; passivity, corrosion or immunity depending on the pH and on the electrical potential of the steel.

Passivity and corrosion are conditions whereby the above reactions are occurring initially but the Fe^{2+} (ferrous) ion is unstable in the presence of oxygen and it will react further forming ferric oxide. The pH and electrical potential will determine the physical nature of the ferric oxide formed and thus whether the iron is corroding or passive.

The corrosion form of iron oxide is expansive, non-adherent and tends to flake leaving the iron surface exposed to further oxidation. At more positive potentials, a passive oxide film is formed on the metal surface. This film is adherent, self-repairing and impermeable, limiting further oxygen access to the metal and effectively halting the corrosion process.

Immunity occurs at more negative potentials than corrosion and is a condition whereby the metal has a bright surface and cannot corrode. So-called 'noble metals' like gold and platinum exist in this condition under normal circumstances but for iron, the electrical potential needs to be so highly negative that it will cause water to break down and bubbles of hydrogen gas will be seen on the surface. This is not the normal condition for steel and will generally only occur where an external potential is applied to it.

The presence of chloride ions significantly modifies the corrosion states of metals including iron. For iron the range of pH and potential for immunity remains the same but the range for general corrosion is considerably extended. In addition, a different type of corrosion called pitting corrosion can occur which is intense and localized. Consequently, the conditions for passivity become severely limited.

7.2.4 Cathodic protection

The idea of cathodic protection is to artificially shift the potential of a metal so that it becomes either immune or passive. In natural soils and waters it is normal to shift the potential of steel to the immune region whereas for steel in concrete it is preferable to re-establish passivity.

In sacrificial anode cathodic protection, a galvanic cell is set up by connecting the steel to a more reactive metal, usually zinc. The zinc then undergoes the anodic reaction and corrodes while the steel is rendered entirely unreactive because the whole surface undergoes the cathodic reaction mentioned above and the iron no longer dissolves. This may also be thought of as the anodic sites on the steel being shifted to the zinc.

With impressed current cathodic protection, the steel is connected to the negative terminal of an electrical power supply forcing it to undergo a cathodic reaction. If the potential of the steel is made negative enough to make it immune, the cathodic reaction becomes one whereby water is broken down and hydrogen is liberated as follows:

$$2H_2O + 2e^- \rightarrow H_2 + 2OH^- \quad (7.3)$$

This situation would normally be avoided in concrete since the pH is high and it is possible to re-establish passivity by applying a somewhat less negative potential. This consumes considerably less current and so reduces the cost. The cathode reaction is then as given in reaction (7.2) above and the OH⁻ generated helps to maintain the conditions necessary for passivity.

The anode, connected to the positive terminal of the power supply, is usually chosen to be a relatively non-reactive conductor such as carbon or titanium so that its corrosion rate is low. The anode reaction then generates oxygen and acid (H⁺) as follows:

$$H_2O \rightarrow O_2 + 4H^+ + 4e^- \quad (7.4)$$

The current densities normally encountered in CP systems are sufficiently low for the amount of acid generated to be safely taken up by the normal alkalinity of the concrete.

7.2.5 Practical anode systems

Techniques for the cathodic protection of steel in concrete vary according to the anode system used. The longest established system uses cast iron primary anodes over a bridge deck with a conductive asphalt secondary anode laid over them and an asphaltic concrete wearing course over that. The system operates at relatively low current densities with an expected life of 10 to 20 years. It is a low-cost, low-technology system which has proved to be highly reliable. Its disadvantage is its limited application being only suitable for tops of decks.

Conductive coating systems typically contain graphite in a resin binder and are applied at a thickness of 250 to 400 microns as a secondary anode. The primary anode consists of thin wires or strips of titanium laid into the surface beneath the coating. The coating itself is black but may painted over. The systems operate at medium current densities with an expected life of 5 to 15 years. An advantage of this type of system is the range of different surface orientations and shapes to which it can be applied.

Expanded titanium mesh may be applied as both primary and secondary anode. This is overlaid often with sprayed concrete, paving concrete or cast superplasticized mortar. Depending on the grade of mesh used, the range of current densities may be very large and design lives of 25 to 100 years or more have been claimed.

The major factor in deciding anode life is the presence of water due to leakage or permeation through the concrete. Attention to detail regarding waterproofing of joints and treated concretes can extend anode life considerably.

Sacrificial flame- or arc-sprayed zinc has also been used successfully used in marine locations. These systems are relatively cheap, require little maintenance and have achieved lifetimes of 8 years or more. An advantage can be that minimal repairs need to be carried out to the deteriorated concrete and, in fact, zinc sprayed over exposed steel is a convenient way of achieving the required electrical contact.

Small sacrificial anodes have been developed to protect steel reinforcement from the incipient anode effect and also to provide sacrificial cathodic protection (Granthen and ross, 2003).

7.3 Electrochemical chloride extraction (desalination) and realkalization

7.3.1 Introduction

This section is concerned with two recently developed processes for dealing with corrosion of steel in concrete. Mechanisms of corrosion of steel in concrete, and other methods of dealing with it are dealt with elsewhere. Although chloride removal (CR, also known as chloride extraction and desalination) and realkalization (ReA) are new techniques, their use will undoubtedly expand with time.

We will first see how CR and ReA stop corrosion. We will then review the advantages and disadvantages of these processes versus other ways of stopping corrosion such as concrete removal, sealers and cathodic protection. We will also look at some case histories of the processes.

The two processes are closely linked, but to avoid confusion this section deals with the generalities of corrosion, and electrochemical techniques and then will deal with CR and ReA separately.

7.3.2 The mechanisms of corrosion of steel in concrete

We already know that steel is normally passive when embedded in concrete, due to the alkalinity of the pore water. There are two ways in which this passivity can be destroyed.

Carbonation

The simplest to understand is by carbonation. Atmospheric CO_2 dissolves in the pore water and forms carbonic acid. As we know, acids react with alkali to form water and a neutral salt, so the carbonic acid reacts with the calcium hydroxide to form calcium carbonate:

$$CO_2 + H_2O \rightarrow H_2CO_3$$

$$H_2CO_3 + Ca(OH)_2 \rightarrow CaCO_3 + 2H_2O$$

Once the calcium hydroxide is consumed the pH drops from 13 to 9 and the passive layer decays. The steel then corrodes in the presence of the oxygen and water available in the concrete *pores*.

Chloride attack

The chloride ion attacks the passive layer, even though there is no drop in pH. Chlorides act as catalysts to corrosion. They are not consumed in the process, but help to break down the passive layer of oxide on the steel and allow the corrosion process to proceed quickly.

There is a well-known 'chloride threshold' for corrosion, given in terms of the chloride/hydroxyl ratio. When the chloride concentration exceeds 0.6 of the hydroxyl concentration, then the passive layer will break down. This *approximates* to a concentration of 0.4 per cent chloride by weight of cement. The approximation is because:

(a) Concrete pH ($14 - \log_{10}$ of the OH concentration), varies with the cement powder and the concrete mix. A tiny pH change is a massive change in OH concentration and therefore the threshold moves radically.
(b) Chlorides can be bound chemically (by aluminates) and physically (by adsorption on the pore walls). This removes them (temporarily or permanently) from the corrosion reaction.
(c) In very dry concrete corrosion may not occur even at very high Cl^- concentration as the water is missing.
(d) In saturated concrete corrosion may not occur even at a very high Cl^- concentration as the oxygen is missing to fuel the corrosion reaction.

Therefore corrosion can be observed at 0.2 per cent chloride, and none seen above 1.0 per cent or more. If (c) or (d) are the reasons, then a change in conditions may lead to corrosion.

7.3.3 Electrochemical processes

Electrochemical techniques include cathodic protection (CP), chloride removal (CR) and realkalization (ReA). They all rely on the fact that the corrosion process is not merely a chemical one, but also involves the movement of electrical charge. When steel in concrete corrodes, it dissolves in the pore water and gives up electrons:

$$Fe \rightarrow Fe^{2+} + 2e^- \quad (A)$$

This is the anodic reaction.

The two electrons must be consumed elsewhere on the steel surface to preserve electrical neutrality:

$$2e^- + 2H_2O + O_2 \rightarrow 4OH^- \quad (C)$$

This is the cathodic reaction. You will notice that we are generating more hydroxyl ions in the cathodic reaction. These ions will strengthen the passive layer, warding off the effects of carbonation and chloride ions. You will also note that we need water and oxygen at the cathode to allow corrosion to occur.

Obviously, if we can make the cathodic reaction predominate along the steel, we will stop corrosion. In electrochemical techniques we apply an external anode to the concrete surface. This generates the electrons instead of the anodic reaction (A), and the steel has only the cathodic reaction (C) occurring on its surface. For CR and ReA, the external anode is temporary, and the reactions are driven by a DC power supply. The systems re-establish a 'passive' environment around the steel that will last many years.

One requirement for all electrochemical treatment, CR, ReA and cathodic protection, is good electrical continuity to ensure that current flows from the anode to all areas of steel. Electrical continuity must be checked and, if necessary, established in all applications of these techniques. The reverse side of that issue is that there must not be short circuits between the steel and the surface. If there is, current will short circuit the concrete pore structure and the chloride ions (for CR) and hydroxyl ions (for ReA) will not flow.

7.3.4 Chloride removal

We have already determined that the chloride ion is a catalyst to corrosion. As it is negatively charged we can use the electrochemical process to repel the chloride ion from the steel surface and move it towards an external anode.

Anode types

The most popular anode is the same coated titanium mesh used for CP. Instead of embedding it permanently in a cementitious overlay, a temporary anode system is used. This is placed inside a cassette shutter, as in Figure 7.1. Where the shape of the member is especially difficult, a sprayed papier-mâché system can be applied over the anode, soaked with the electrolyte. Systems have also been developed that use an ion-exchange resin to trap the chloride extracted from the structure (Schneck, 2003).

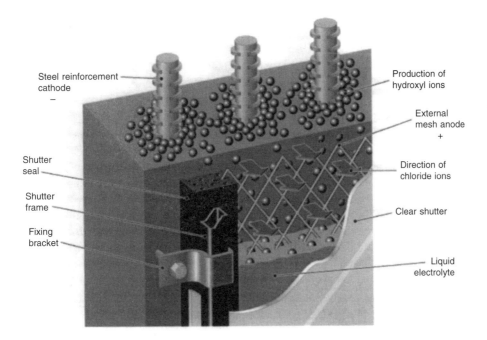

Figure 7.1 Schematic arrangement showing desalination.

A 'sacrificial' anode can be used instead of coated titanium. Copper was tried in the early 1970s trials, but copper (or its salts) may accelerate corrosion if it gets into concrete. Steel mesh has been used more recently but has fallen out of favour as it may be completely consumed in some areas before the treatment process is over.

Electrolytes

Calcium hydroxide is the usual electrolyte. This has the advantage over water of being alkaline and stops chlorine gas evolution, although additional chemicals may be added to increase alkalinity. Lithium salts have also been used to stop ASR (see below).

Operating conditions

The very first trials of this system were based on a rapid treatment period of 12–24 hours. Trials were done in Ohio USA, and lab tests in Kansas Department of Transportation in 1978. However, a more fundamental study carried out by the Strategic Highway Research Program (SHRP) in 1987–92 showed dangers in applying more than about 2 A/m² of steel or concrete surface area.

The Ontario Ministry of Transportation carried out a trial on Burlington Skyway in 1989. The voltage was kept at about 40 V. The total charge passed was 610 Ah/m² of concrete surface area over 55 days giving an average current density of 0.462 A/m². This structure had a low steel concentration, so removal was patchy, very high over the rebars, but lower between the bars. They found 78–87 per cent of the chloride removed directly above the rebars and 42–77 per cent of the chloride removed between the rebars.

A trial was carried out in the UK on a section of crosshead taken from a corroding substructure. The section was removed to a contractor's depot and the sprayed cellulose fibre system applied. The system ran for 92 days passing a total of about 19 565 amp hours charge through approximately 11 m² of steel surface, giving a charge density of 1704 A.h/m², an average current density of 0.77 A/m² and a power density of 25 W/m².

End point determination

End point determination can be by several means:

(a) Point of diminishing returns – resistance goes up, amount of chloride removed goes down, when the current is small and the amount of chloride removed is small, switch off. Switching off for about a week will bring the system resistance down. But how much more chloride is removed by allowing 'rest' periods is not known.
(b) Direct measurement – take samples from the concrete and when an agreed level is reached, stop. This assumes that good sampling is possible and that samples are representative.
(c) Indirect measurement – sample the anode system and/or electrolyte. When chloride level is either at a plateau or an agreed level, stop. This assumes good sampling.
(d) Experience of charge density needed – measure charge passed (amp hours per square metre) and when an agreed limit is reached, switch off.

In practice a combination of systems is used. The experienced contractor will know the charge density needed and a trial may give a definite value for a given structure (or element within the structure). Sampling directly and indirectly will show that the system is responding in the expected manner. The point of diminishing returns should be reached soon after the other thresholds.

It is of course impossible to remove all the chlorides from the concrete. The area immediately around the rebar is almost chloride free, but further away there is less effect. This is particularly true behind the steel. Chloride removal will deplete the amount of chloride immediately in contact with the steel, and will replenish the passive layer. Field data shows that this is effective for at least six years, but for how much longer is uncertain. The results suggest about 10 years, but only real experience will show that.

One implication of the amount of chloride which can be removed is that if large amounts of chloride have penetrated beyond the steel, or were cast uniformly into the concrete, then chloride removal will only affect the chloride level in the 'covercrete'. The chlorides in the bulk of the concrete will then diffuse back around the steel and the corrosion may eventually be re-established. When the process is carried out, the steel is polarized for some time, perhaps a year, making half cell potential testing difficult to interpret. The charge on the steel means, however, that chloride ions will be repelled by the charge during this time. Thereafter, there will still be a reservoir of negatively charged hydroxyl ions around the steel, and work is currently being undertaken to define how long this additional protection remains in place.

Possible effects

Passing large amounts of electricity through concrete can have effects upon its chemistry and therefore its physical condition. Brown staining around the rebar has been observed on specimens when high currents and voltages are used (in excess of about 2 or 3 $A.m^2$). This is an effect on the concrete, not the steel. Current levels are therefore carefully controlled using an expert management system, with data logging equipment built into the power supply.

There are two known side effects of CR. The first is the acceleration of alkali–silica reactivity (ASR), and the other is reduction in bond at the steel concrete interface.

Alkali–silica reactivity

Research at Aston University in England and by Eltech Research in the USA under the SHRP programme shows that there can be excess ASR induced by the cathodic reaction (C) that generates excess alkali at the steel surface. This is exacerbated by the movement of alkali metal ions (Na^+ and K^+) to the steel surface under the influence of its negative charge. Some researchers in Japan have suggested that the pH can be so high that the silica gel dissolves, stopping the expansive process.

Eltech have undertaken field trials to see if that can be controlled by the application of lithium ions in the electrolyte. Lithium is known to reduce or stop ASR, and proved effective in lab tests. If a corroding structure is made with aggregates susceptible to ASR, a detailed investigation of its likely reaction to CR will be required.

Bond strength

The effect of current on bond strength of steel in concrete has been a subject for discussion in the literature for many years, usually with reference to cathodic protection. In most practical applications, the major part of the bond is supplied by the ribbing on the bars, so the details of the performance of the steel/concrete interface is irrelevant.

A feasibility study is presently underway for the application of CR to a long bridge structure in the North of England. In this case, extensive use was made of smooth rebars, due to steel shortages at the time of construction. Some tests have shown that CR reduces bond strength by as much as 50 per cent. However, careful review of the laboratory data shows that a large amount of charge was passed to achieve this drop. About five times as much as is used normally in a CR treatment. A very high current density was also used.

Work has been carried out by Imperial College that showed reduction in bond could be a problem with smooth bars, although the effects were small and in no case did the bond drop below that of the control, uncorroded, sample. It was considered likely that the lowering of bond was due to passivation of the corrosion deposits. In any case, spalled concrete, which is the other alternative, is not noted for its bond to steel reinforcement! Ribbed bars showed no reduction in bond strength.

Case histories
Tees

The A19 Tees viaduct in the UK was opened to traffic in 1975. Major refurbishment work on the bearings and deck was started at the end of 1987. At the end of 1991, work started on the problem of the corroded reinforcement in the piers that support the deck. There are approximately 70 piers. About 37 were candidates for chloride removal or similar refurbishment. A further eight could have been included. Another 12 or so had to be removed, two having been replaced already.

When the first pier was demolished, a central section was cut out whole and shipped to a contractor who conducted a chloride-removal trial. Wooden battens were nailed to the surface, a layer of cellulose fibre was sprayed onto the surface, and the coated titanium mesh anode nailed to the battens. A final layer of fibre was then applied.

The system ran for 92 days passing a total of about 19 565 amp hours charge through approximately 11 m^2 of steel surface, giving a charge density of 1704 $A/h.m^2$, an average current density of 0.77 A/m^2 and a power density of 25 W/m^2. This trial culminated in the work undertaken at Imperial College, mentioned above.

Chloride levels dropped from an average of 1.13 per cent chloride (standard deviation 0.96 per cent) to 0.25 per cent (standard deviation 0.16 per cent). Potentials (based on a copper/copper sulfate half cell, but ignoring the negative sign) were all below 200 mV, having started in the range 100 to 600 mV. Chloride removal was the preferred method of protection on this structure due to the public access to the piers that could lead to vandalism of a cathodic protection installation. Chloride levels were so high in the piers that concrete removal and repair might not have been economic. The only alternative would be replacement of the piers, undertaken using a special temporary works structure with hydraulic jacks to support the deck while the pier is demolished and rebuilt.

Burlington

Burlington Skyway is a major flyover on the Queen Elizabeth Freeway in Ontario, Canada. It was constructed in 1955 and has suffered from chloride-induced corrosion over many years. It has been a trial site for several cathodic protection systems. Dr David Manning of Ontario Ministry of Transportation decided to use this site for a trial of CR when the commercial system was licensed for use in North America.

The principal difference between this and the Tees trial was the use of a mild steel mesh that corroded away sacrificially as the process proceeded. This left rust stains on the concrete that were grit blasted off when the trial was finished.

Three sides of a column were used for the trial. The fourth side was monitored as a control. The steel density was low (0.55 m^2 of steel per 1 m^2 of concrete surface on the east and west sides and 0.79 m^2/m^2 on the north and south sides). The voltage was kept at about 40 V. The total charge passed was 610 Ah/m^2 of concrete surface area over 55 days giving an average current density of 0.462 A/m^2 concrete surface area. This is approximately 900 Ah/m^2 of steel surface area, and about 0.7 A per m^2. The low steel concentration meant that removal was patchy, very high over the rebars, but lower between the bars. They found 78–87 per cent of the chloride removed directly above the rebars and 42–77 per cent of the chloride removed between the rebars. Latest results show the steel passive and very low half cell potentials. The trial was conducted in July and August 1989.

Why choose chloride removal?

One of the major issues facing any consultant or owner of a structure suffering from chloride-induced corrosion is what form of repair to undertake. There are coatings and sealants, specialized patch repair materials, options for total or partial replacement, cathodic protection and now chloride removal.

Since the pores in concrete contain significant amounts of water and air, sealing it to stop corrosion is unlikely to be effective once chlorides have penetrated the concrete. This comparatively cheap solution is not effective once corrosion has started.

Patch repairs are only effective if chloride ingress is local and the chlorides can all be removed. Just patching up damaged areas is a very short-term palliative (owing to the incipient anode effect), not a long-term rehabilitation. Contaminated concrete must be removed from all around the steel, and support may be needed during the repair process.

Cathodic protection has been described as the only solution to chloride-induced corrosion that can stop (or effectively stop) the corrosion process. It has been applied to concrete bridges in the USA since 1973, and in the UK to buildings and other structures since 1988. The problems with cathodic protection are its requirement for a permanent power supply, regular monitoring and maintenance.

Chloride removal has the advantage that, like a patch repair, it is a one-off treatment. A generator can be brought in for the duration of the treatment, so mains power is not needed. There is no long-term maintenance need, but the system does treat the whole structure.

The disadvantage is its unknown duration of effectiveness. We cannot remove all the chlorides from the concrete. If we can stop further chloride ingress then the system may be effective for many years (10 to 20). If chlorides are still impinging on the structure then it will be shorter (or may require a coating or sealant).

Chloride removal can only be applied to pre-stressed structures with great care due to the risk of hydrogen embrittlement. Work is progressing on its application to structures suffering from ASR, but this is in the early development stage at the moment. As stated earlier, there must be electrical continuity within the reinforcement network for either of these techniques to be applied.

Conclusions about chloride removal

The origins of chloride removal lie in trials carried out in the 1970s. It is now being vigorously pursued by the developers of the Norwegian system, based on the spayed cellulose anode and a major US research programme. The Ontario Ministry of Transportation now has three trial systems (two of the Norwegian system, one of the US SHRP researchers) applied. There is a trial and ongoing monitoring underway in the UK. Results are all encouraging.

We do not know how long the treatment process will last, but a range of five to twenty years is likely, depending upon conditions.

7.3.5 Realkalization

In equations (7.1) and (7.2) we saw how carbonic acid reacts with calcium hydroxide to form calcium carbonate. This removes the hydroxyl ions from solution, and the pH drops, so that the passive layer is no longer maintained and corrosion can be initiated.

The cathodic reaction (C) showed that by applying electrons to the steel, we can generate new hydroxyl ions at the steel surface, regenerating the alkalinity, and pushing the pH back.

The realkalization process has been patented, and uses the same cassette shutter or sprayed cellulose system developed to apply chloride removal. In addition to generating hydroxyl ions, the developers claim that by using a sodium or potassium carbonate electrolyte they make the treatment more resistant to further carbonation.

The patent claims that sodium or potassium carbonate will move into the concrete under electro-osmotic pressure. A certain amount will then react with further incoming carbon dioxide. The equilibrium is at 12.2 per cent of 1 M sodium carbonate under atmospheric conditions.

$$Na_2CO_3 + CO_2 + H_2O \rightarrow 2NaHCO_3$$

Laboratory tests have shown that it is very difficult, if not impossible, for a treated specimen to carbonate again. Over 80 realkalization treatments (50 000 m^2) have been undertaken on structures around Europe over the past few years. The treatment is faster than chloride removal, only requiring a few days of treatment. Work by Banfill at Heriot Watt University (Al Khadimi et al., 1996) showed that realkalization actually improved the properties of concrete, reducing porosity and permeability, increasing strength and modulus and not apparently causing any difficulties.

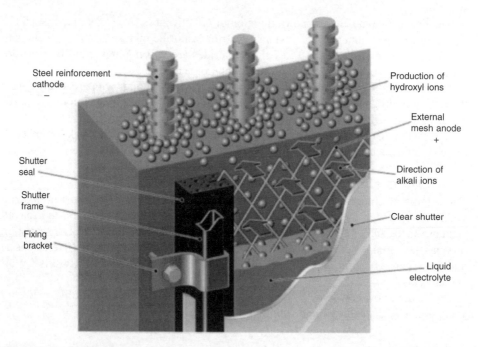

Figure 7.2 Realkinization.

Anode types

Anode types are the same as for chloride removal. Cassette shutters or sprayed cellulose are used by the owners of the patented system, with a steel or coated titanium mesh. The steel is more likely to be used here as the treatment time is shorter and the steel is less likely to be completely consumed.

Electrolytes
As stated above, sodium carbonate solution is the preferred electrolyte to give long-lasting protection against further CO_2 ingress. However, introducing sodium ions can accelerate ASR so in some cases just water is used. Also, some problems have been found with adhesion of coatings when sodium carbonate is used, so the tendency is now to use a very low dose of sodium carbonate, or just to use water.

Operating conditions
In one case a current density of 0.3–0.5 A per m^2 was applied (at 12 V) to 2000 m^2 of a building in Norway, with a treatment time of 3–5 days. In another case 10–22 V was applied to give a current density of 0.4–1.5 A/m^2 in 12 days on 300 m^2 of a bridge control tower in Belgium. A further section of 140 m^2 was treated in 9 days with a current of 1–2 A/m^2. All figures are for concrete surface area. The steel to concrete surface ratio was not given.

End point determination
This is easy for carbonation. A simple measurement of carbonation depth will show when it has been reduced to zero. Measurement of current flow also gives guidance and a rapid on-site method for measuring sodium content of dust samples has been developed, where sodium carbonate is used.

Possible effects
As there is a smaller charge density applied, the risks of damage are lower than for chloride removal. As mentioned above, ASR is a risk if sodium carbonate is used as the electrolyte.

Sodium carbonate can also cause short-term efflorescence, and the high alkalinity after treatment can attack some coatings. Sodium carbonate will attack oil-based paints, varnishes and natural wood finishes.

Case histories
The contractors who own the patents claim that over 80 structures have been treated with realkalization around Europe. Two cases are summarized above. A recent application was completed on 1500 m^2 roof area of Walthamstow Magistrates Court. Carbonation depths ranged from 5–25 mm. After treatment a polymer-modified mortar was applied to the surface and an elastomeric decorative finish applied. The advantage to the client in this case was the lack of noise, so the court could remain in session.

Why choose realkalization?
Realkalization is a simpler treatment than chloride removal. However, the alternative of patching and coating with an anti-carbonation coating is much more effective than patching and coating for chloride attack. The extent to which carbonation has reached the rebar and the requirements for patch repairing to restore alkalinity will determine whether realkalization is preferred either because it is more economic, or in other cases to avoid the noise, dust and vibration required for extensive patch repairing.

The technique is becoming popular in Europe and the Middle East (in North America very little attention is paid to carbonation). If the treatment is as effective as claimed, then the choice between realkalization and patching and coating is a question of convenience and cost, together with a realistic appraisal of the effectiveness of anti-carbonation coatings. The patentees and the licensees of the system claim that:

(a) It is financially competitive with the alternatives.
(b) There is greatly reduced vibration and noise.
(c) All of the surface is treated.
(d) Guarantees are offered.

7.3.6 Conclusion

There is no doubt that both these techniques will become established in the concrete repair field. They have some unique advantages which will make them the system of choice for some applications. For other applications they will need to compete financially with other systems.

7.4 Corrosion inhibitors

These materials have seen a dramatic increase in promotion via manufacturers such as Sika, Balvac, Grace and others. They vary in mode of use. There are two main types:

(a) Penetrating liquid or vapour phase corrosion inhibitors (Sika Ferroguard, Balvac MFP)
(b) Concrete additives such as calcium nitrite, designed to be added to the concrete at the time of placing, perhaps in conjunction with ggbs, pfa or silica fume.

The latter have a good track record of successful use in the USA, employ sound technology, and have been shown by trials to be effective at dealing with chloride-induced corrosion for external sources of chloride, such as de-icing salt.

The former are materials such as amino-alcohols (Ferroguard), or sodium monofluorophosphate (MFP). These material are claimed to deal with both carbonation-induced corrosion, and relatively high levels of chloride. The technology on which they are based has been successfully used in the metals industry for many years. They work by inhibiting both the anodic and the cathodic corrosion reactions, by penetrating in the liquid or vapour phase down to the steel reinforcement and surrounding the steel.

Research by various workers has been carried out and it is evident that in some low-risk situations, these materials can have a role in concrete repair. Based upon current research, their effectiveness is limited to

1 Concrete suffering from carbonation with shallow cover, say up to 15 mm. Their ability to work with deeper cover depends on their ability to penetrate adequately to the reinforcement. This varies from concrete to concrete and cannot be guaranteed.
2 Concrete containing up to 0.6 per cent chloride as chloride ion, by mass of cement. Research has not supported their effectiveness at higher values.

Notwithstanding this, repair contracts are being carried out using a combination of patch repairs, corrosion inhibitors and coatings. In some cases an insurance-backed warranty is being offered, for typically 10 years. Given the combined effectiveness of inhibitors and coatings together, with existing damage dealt with by patch repair, then if the warranty is considered acceptable, we see no reason why the client should not accept this approach, provided caution is exercised. We would not, however, consider this approach to be likely to be successful in high-chloride situations. At chloride levels above 1 per cent, specific reassurances should be sought, in our opinion.

Acknowledgements

The author would like to acknowledge the help of Dr John Broomfield in preparing this chapter. Data from Dr David Manning of the Ontario Ministry of Transportation is also gratefully acknowledged, as well as background data supplied through John Broomfield's consultancy work with the Strategic Highway Research Program.

References

Al. Khadimi, T.K.H., Banfill, P.F.G. *et al.* (1996) An experimental investigation into the effects of electrochemical re-alkalization on concrete. *Proceedings of an International Symposium on the Corrosion of Reinforcement in Concrete Construction*, Cambridge, 1996. Royal Society of Chemistry.

Anderson, G. (1992) Chloride extraction and realkalization of concrete. *Hong Kong Contractor*, 19–25, July–August.

Bennett, J.E. and Schue, T.J. (1990) Electrochemical chloride removal from concrete: a SHRP contract status report. Corrosion 90. April 23–37: Paper 316.

Broomfield, J.P. (1990) *SHRP Structures Research*. Institute of Civil Engineers ICE/SHRP, Sharing the Benefits; 29–31 October 1990; Tara Hotel, Kensington, London. London: Imprint, Hitchin, Herts; 1990; ICE 1990: 35–46.

Concrete Society (1991) Patch repair of reinforced concrete, subject to reinforcement corrosion. Concrete Society Technical Report No. 38. Report of a Concrete Society Working Party.

Electrochemical Chloride Removal and Protection of Concrete Bridge Components: Laboratory Studies (1993) Strategic Highway Research Program Report SHRP-S-657, National Research Council, Washington, DC.

Grantham, M. and Ross, R. Repairing concrete car parks – a case study from the clients' perspective. *Proceedings of 'Concrete Solutions' 1st International Conference on Concrete Repair*, St Malo, France, July 2003.

Jayprakash, G.P., Bukovatz, J.E., Ramamurti, K. and Gilliland, W.J. (1982) *Electro-osmotic techniques for removal of chloride from concrete and for emplacement of concrete sealants*. Kansas DoT Final Report; August 1982; FHWA-KS-82-2.

Lankard, D.K., Slater, J.E., Diegle, K.B., Martin, C.J., Boyd, W.K. and Snyder, M.J. (1978) *Development of Electrochemical Techniques for Removal of Chlorides from Concrete Bridge Decks*. Battelle, June 1978; Revised Final Report.

Lankard, D.R., Slater, J.E., Diegle, R.B. and Boyd, W. K. (1978) *The Electrochemical Removal of Chlorides from Concrete: Guide to Materials and Methodology.* Battelle, June.

Lankard, D.K., Slater, J.E., Redden, W.A. and Neisz, D.E. (1975) *Neutralization of Chlorides in Concrete*, FHWA-RD 76-60 September.

Miller, J.B. (1989) *Chloride Removal and Corrosion Protection of Reinforced Concrete*. Strategic Highway Research Program and Traffic Safety on Two Continents, Gothenburg Sweden. 29, September.

Schneek, U. *et al.* Electrochemical chloride extraction from a pre-stressed box girder floor slab as an alternative to complete replacement – a case study. *Proceedings of 'Concrete Solutions' 1st International Conference on Concrete Repairs*, St Malo, France, July 2003.

PART 3

Quality and standards

8

Quality concepts

Patrick Titman

8.1 Introduction

The ACT syllabus embodies the truth that technical knowledge and expertise are of limited value unless they are linked to management expertise. Part of the approach to this expertise is an understanding of the various management systems expected to operate – quality, environment, health and safety – and of how these systems apply in different parts of the construction industry. For a firm to be successful it must have people able to be sympathetic to and to relate the firm to the needs of its customers. This chapter addresses these issues.

8.2 Definitions

Formal definitions of quality management terms are given in ISO 9000: 2000 *Quality management systems – fundamentals and vocabulary*. These can be expressed in the simple terms below:

Quality	Meeting requirements.
Quality management	The means we devise for meeting requirements.
Quality control	The means of knowing (monitoring) where we are relative to meeting requirements.
Quality assurance	The assurance we give ourselves and others – customers, shareholders, regulators – that quality is being provided.
Audit	Independent enquiry to examine whether the management is appropriate and whether quality is being achieved.

8/4 Quality concepts

But what is really useful is to understand the relationship between the five concepts. They make a pattern like the clock in Figure 8.1. Quality is the right time. The face of the clock shows this and gives assurance to the viewer that the time is told. The works are the management that strives to make the face right. The part of the works – in a clock the regulator that monitors whether the time is right – is the quality control function. Audit is outside the clock, it takes a cool view as to whether it is an effective working clock, telling the right time, or whether it needs the oil of improvement and the regulator adjusting.

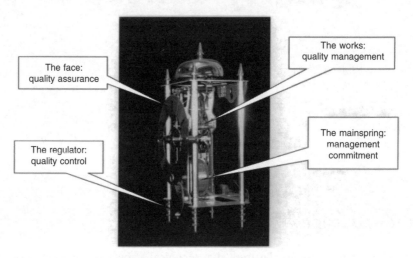

Figure 8.1 Quality definitions: the quality clock.

Once the idea that quality is meeting requirements is accepted, it is easy to think of a whole range of requirements, not just the customer's requirement for a particular product such as concrete of a defined strength and durability, but other requirements such as the customer feeling satisfied; safe production of concrete with minimal risks to health; production in an environmentally beneficial way; concrete manufacture that creates profit at a planned rate to satisfy shareholders and so on. It is an insight of the total quality management movement that allows this extension of the application of quality systems to all spheres of a company's business. These requirements often appear to conflict. It is satisfactory systems management – coordinating each aspect of management – that allows practical resolution of such conflicts.

One other term needs explanation. It is 'arrangements'. Health and safety literature talks of arrangements rather than management system. The two overlap: management systems look to overall management to provide an environment in which planning ahead leads to satisfactory outcomes and review leads to improvement. Arrangements look to practical instructions and resources so that satisfactory outcomes can happen and feedback can drive improvement. But the terms are two halves of the same coin: the practical arrangements will not be in place consistently unless there is overall management to direct and secure resources.

8.3 Systems management standards

8.3.1 The primacy of ISO 9001

The principal member of the family of management standards is ISO 9001: 2000. It is the standard with the longest history. Other standards depend upon it for framework and good practice. Understanding it opens the way to considering the family of management standards as a whole.

8.3.2 Understanding the ideas of ISO 9001: 2000

ISO 9001 is founded on two concepts and eight management principles. The concepts are:

1. What organizations do is to be regarded as a series of interlinked and/or nesting processes.
Each process is modelled by Figure 8.2.

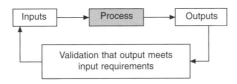

Figure 8.2 The process model.

Interlinked processes can arise as shown in Figure 8.3 where an output of one process is an input of the next.

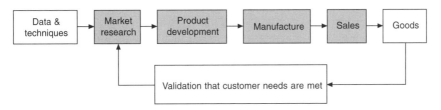

Figure 8.3 Interlinked processes.

Nesting processes arise when a major process is seen as containing a series of sub-processes e.g. within 'manufacture' there are a series of sub-processes such as 'purchasing', 'assembly', 'inspection' and 'calibration'.

There are only a limited number of inputs – people, information, plant and equipment, materials, finance, a workplace such as an office, factory or site. Outputs can be characterized as those for the customer, those for the public, those for members of the organization and those for the organization's owners. Some outputs are tangible, e.g. a manufactured item, others are not, e.g. enhanced reputation. Less tangible things stand behind some inputs too, e.g. motivation of people, confidence of financial bodies.

2. Each process is to be tested against the cycle of activities in Figure 8.4.

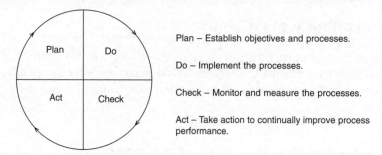

Figure 8.4 The continuous improvement cycle (after Deming).

The eight management principles are intended for use by top management to lead the organization towards improved performance.

The principles are referred to in section 0.1 of the standard, but they are only listed out in the guidance standard (ISO 9004: 2000 cl. 4.3). The principles require management to display:

- Customer-focused organization.
- Leadership.
- Involvement of people.
- Process approach.
- System approach to management.
- Continual improvement.
- Factual approach to decision making.
- Mutually beneficial supplier relationships.

Application of these principles is easiest to envisage at the highest level in an organization and the standard makes much of the need for 'top management' to participate in giving substance and authority to any system generated to meet it. At this level the standard owes much to the ideas of corporate planning (see e.g. Argenti, 1989). Evidence of application of most aspects of the principles will be found in the plans, shown in Figure 8.5, commonly generated by the corporate planning process.

Other aspects – leadership, involvement of people, have to be found in how these plans are communicated and embraced throughout the organization and the extent to which comment and discussion are able to influence them. Communication is not just with members of the organization: it includes customers and suppliers. It also includes shareholders and the public though this is less clearly identified in the standard.

The wording of ISO 9001: 2000 concentrates on the use of the principles by top management. But while 'top management' means 'those who direct and control an organization at the highest level' (ISO 9000: 2000 cl. 3.2.7) it is an elastic phrase when related to location: in construction particularly, the local top manager is whoever is in charge at the site. Moreover, tomorrow's top managers are today's junior ones. In any event, genuine application of the principles will be visible at every level in the organization – management, supervisors and workpeople (see Figure 8.6).

It is impossible to look at a quality system separate from the business management system as a whole. Corporate planning balances the need for commercial achievement

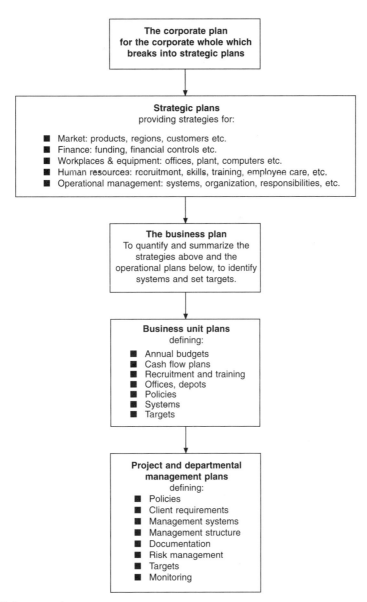

Figure 8.5 Corporate plans.

against the quality and reliability of product or service needed to realize this. Both exist in societies where training, health, safety and environmental concerns have to be satisfied, partly as a matter of law and partly to assist in motivating employees, customers and the public. The current edition of ISO 9001 recognizes these interactions to a greater extent than previous editions, but to realize them to the full it's necessary to look at the standards – such as ISO 14001 for environment – describing these systems in their own right and, as more and more organizations are finding, to write a single management system, coordinating all the aspects above. It is only when this happens that there is a satisfactory

8/8 Quality concepts

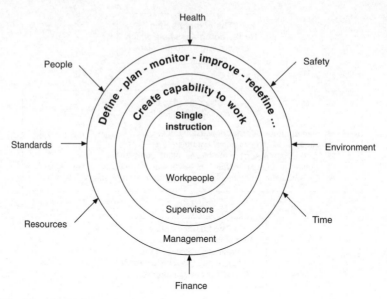

Figure 8.6 The integrated management approach.

and overt system to balance the sometimes conflicting pressures on the people of the organization, resulting in clear and unequivocal instructions for doing work (see Figure 8.6).

8.3.3 Understanding the text of ISO 9001: 2001

ISO 9001 is structured in eight main clauses (sections), with an additional introduction. The introduction identifies the two concepts set out above and sets out briefly the relationship with other standards and management systems. The first three main sections deal with scope, application and definitions. Points of interest in these sections are:

- (Scope) Making an aim to demonstrate that customer and regulatory requirements are met while omitting any reference to other stakeholder (e.g. shareholder) requirements.
- (Application) The permission to exclude irrelevant requirements, but only from among those in section 7, and still claim compliance with the standard. This is primarily to allow design to be excluded but could be applied to other clauses where these do not affect the ability to provide product that meets customer and regulatory requirements.
- (Definitions) A general reference to the vocabulary set out in ISO 9000: 2000 and the particular change from the 1994 standard that 'supplier' now means all, including subcontractors, who supply the 'organization', that is, the body applying the standard to itself and 'customer' means the recipient of the goods or services from the organization. These terms are summarized as follows:

$$\text{Supplier} > \text{organization} > \text{customer}$$

The remaining sections form the bulk of the standard and are headed:
4. Quality management system.
   General, documentation requirements.

5. Management responsibility.
   Commitment, customer focus, policy, planning, responsibility authority and communication, review.
6. Resource management.
   Provision, human resources, infrastructure, work environment.
7. Realisation.
   Planning, customer-related processes, design, purchasing, production [service provision], control of monitoring and measuring devices [calibration].
8. Measurement, analysis and improvement.
   Monitoring and measurement, nonconformity, analysis, audit improvement.

Section 4 lays out the general form of system, but needs to be read with the requirements for policy content in section 5.3 and the requirements spread through the standard for mandatory written procedures. Six of these are identified: document control, records management, audit, identification and control of nonconforming product, corrective and preventive actions. The general form of documentation that emerges in systems meeting any of the management standards is shown in Figure 8.7.

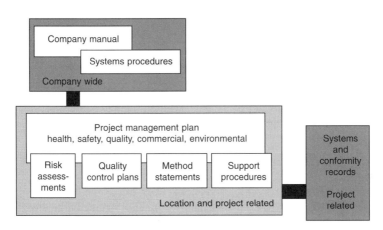

Figure 8.7 Typical systems documentation.

Each paragraph of section 5 begins 'Top management shall . . .'. The section sets out what is required of top management in setting up and maintaining a system. The section is awkward because it only deals with requirements for the system and for seeing that customer requirements are met. Yet all top managers have to be concerned with matters of efficiency and return on capital employed. A more broadly based approach is useful in practice.

Section 6 is concerned with what has to be provided for work to be done. It is broadly based and offers the opportunity to bring together the various systems needed for the organization to function effectively.

Section 7 is very much the same as section 4 of the 1994 standard. It provides the details of the system required to do successful work in practice.

Section 8 is to be applied to sections 4 to 8 inclusive. It provides a methodology for improvement comprising internal audit, monitoring and measurement, handling nonconformity, analysis, preventive action. These techniques are to be applied to product or service realization, to customer satisfaction and to each aspect of management – system, responsibilities, resources and measuring/monitoring techniques. The omission of a requirement to assess the satisfaction of other stakeholders is again noticeable.

8.3.4 Reconciling ideas and text

The idea of process is readily compatible with the text of the standard. Modelling the operations of the organization as a set of processes with listed inputs and outputs provides a convenient way of seeing whether the planning of product realization (cl. 7.1) and of design (cl. 7.3.1) meet the standard and give an acceptably low risk of requirements not being met. Figure 8.8 illustrates this for a civil engineering design office – note that the italic comments suggest possible improvements. Similarly, it is easy to apply the cycle of plan, do, check, act to any process and so to see whether section 8 of the standard is being applied in a useful way.

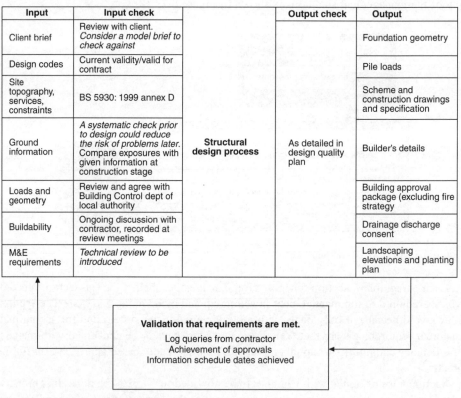

Figure 8.8 Example process model for structural design.

It is a little more difficult to relate the eight principles to the text. This needs to be done, as suggested above, for three levels in the organization – top management, the executive and the workpeople. A helpful way of doing it is to relate each principle to relevant clauses in the text, using the resulting matrix, shown in Table 8.1, as a checklist that each principle is addressed at each level.

Table 8.1 Relationship between principles and clauses of ISO 9001: 2000

Principle	Relevant clause in ISO 9001: 2000		
Management level	Management	Supervisors	Workpeople
Customer focus	5.1a, 5.2, 5.5.2c	7.2, 8.2.1	7.3.3, 7.5.4, 7.5.5, 8.3, 8.5.2, 8.5.3
Leadership	5.1, 5.4.1	6.2.2d	
Involvement of people	5.5.1, 5.5.3, 6.2	6.2, 7.2.3, 7.3.1 para 3 7.3.4 last para	8.2.1, 8.2.2
Process approach	4.1a & b, 5.4.2	7.1	7.3.2 to 7.3.5 inc, 7.5.1
System approach to management	4.1, 5.1, 5.4.2	6.3, 6.4	7.5.1
Continual improvement	5.6.1	8.4, 8.5.1	8.5
Factual approach to decision making	5.6	5.6, 8.1	8.2.3, 8.2.4, 8.4
Mutually beneficial supplier relationships	5.3a, 5.4.1, 5.6	7.4, 8.4d, 8.5.1	

Constructing this table identifies that the text in the standard doesn't always make the relationships obvious. Particular points to remark on are:

- **Customer focus** Customer focus cannot be achieved and maintained in an organization without ongoing research and communication of the results to employees at all levels. It is all too easy to imagine what customers want without testing whether they do. It's also easy to make customer focus no more than persuading customers to accept, not what is best for them but what is best for the organization. If this is to be avoided conflicts of interest such as that shown in Figure 8.9 have to be resolved. This may seem easy – the short-term interest of the constructor (speed) can be reconciled to the longer-term interest of clients (durability) by considering the effect of early failures or dissatisfaction on future market share. Yet this argument is only powerful where the organization sees a long-term future and where employees see continuity of enjoyable and well-rewarded employment.

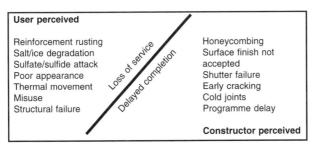

Figure 8.9 User and constructor views: perceived risks to concrete.

- **Involvement of people** Although the standard deals clearly with communication downwards from top management, its very emphasis on top management initiative detracts from strong upward input and from genuine dialogue between different levels of the organization. The idea of communication as a two-way process is better developed in OHSAS 18001: 1999 *Occupational health and safety management systems – specification.* It's worth quoting the relevant clause in full:

> *Consultation and communication*
> *The organization shall have procedures for ensuring that pertinent OH&S information is communicated to and from employees and other interested parties.*
> *Employee involvement and consultation arrangements shall be documented and interested parties informed.*
> *Employees shall be:*
> – *involved in the development and review of policies and procedures to manage risks;*
> – *consulted where there are any changes that affect workplace health and safety;*
> – *represented on health and safety matters; and*
> – *informed as to who is their employee OH&S representative(s) and specified management appointee (see **4.4.1**).*

It's to be hoped that these ideas will come into ISO 9001 at the next revision. In the meantime, good organizations will give effect to them anyway, so allowing workpeople to contribute more positively to adopting the principles in practice.

- **Mutually beneficial supplier relationships**   This principle could be thought to have been abandoned in the practical clauses of the standard: they are all phrased in terms solely of benefit to the organization. It is only within the framework of an overall policy declaration that ideas of mutual benefit between a firm and its suppliers can be seen to be implied.
- **Factual approach to decision making**   The techniques of total quality management (BS 7850: Part 2: 1994) – benchmarking, brainstorming, control charts, Pareto diagrams etc. are relevant here. They aren't referenced, but in the guidance standard quality cost analysis is and self-assessment leading to benchmarking is spelt out in detail for analysis of the quality system.

The whole section falls short of practicality because it is nowhere linked to risk management. Figure 8.10 gives a listing of commonly perceived risks to concrete, compiled from work at BRE and elsewhere. A management system that fails to give priority to addressing these topics will fall short of being effective and there is guidance outside ISO 9001 about how to go about this.

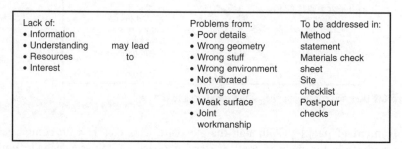

Figure 8.10 Factual approach: risks to concrete.

BSI published BS 6079-3 in 2000 that addresses the topic of risk management and, again, it is to be hoped that the next edition of ISO 9001 will incorporate these ideas. They are already well dealt with in the environmental, health and safety standards. For the present, the good organization will see that risk assessment is used as a tool in analysing data under section 8.4 and that decisions are taken after consideration of the analyses. Figure 8.11 details the risk management process setting the familiar health and safety scheme against that from BS 6079-3. The BS 6079 chart can be seen to relate to each level of organizational planning, aligning with the corporate planning structure shown in Figure 8.5. Both can be useful to use in assessing the quality of analysis under ISO 9001 cl. 8.4 and of decision making under cls 5.6 and 7.2.2c.

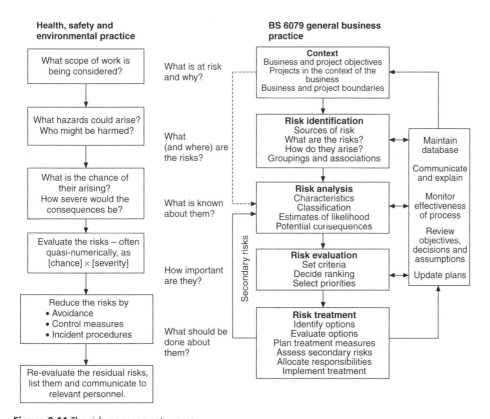

Figure 8.11 The risk management process.

8.3.5 Procedures and method statements

As mentioned above, ISO 9001 makes only six procedures mandatory. Only one of these – handling nonconformity – is concerned with product, the others deal with administration. Other procedures may be introduced and these need not necessarily be written: it may be that training + a prescribed record form, or just a chart on the wall are sufficient. Nonetheless, for carrying out work, cl. 5.5.1 requires responsibilities to be defined, cl. 7.5 points to the possibility of 'work instructions' being needed and gives an indication that these are to deal with information, equipment, monitoring, measuring and release/acceptance. The next clause identifies the likely need for procedures to define product validation. Within civil engineering these matters are universally and fully dealt with in method statements. The contents can be summarized as:

Management support + Methods + Quality control: inspection & test

- Scope and references
- Responsibilities & authority
- Document handling
- Dealing with problems
- Training and qualifications
- Audit
- Communication

- Resources: personnel, plant, materials
- Programme and work methods
- Health and safety precautions
- Environmental practice
- Incident procedures

- Condition at start
- Sampling method and frequency
- Methods & equipment
- What is acceptable
- Reporting
- Keeping/throwing away samples
- Analysis

8.3.6 The family of systems management standards

The family of standards comprises:

Vocabulary
Quality ISO 9000: 2000.
Environment ISO 14050: 1998

Systems standards
Quality ISO 9001: 2000
Environmental ISO 14001: 1996
Health and safety OHSAS 18001:1999

Guidance to systems standards
ISO 9004: 2000 [includes text of ISO 9001]
ISO 14004: 1996

Standards for supporting processes
Auditing: ISO 19011: 2002 'Guidance for quality and environmental management systems auditing'.
This is the most important supporting standard.

Calibration and control of measuring equipment ISO 10012 [various parts]
Not often relevant but referenced in ISO 9001: 2000 cl. 7.6 'Control of monitoring and measuring devices'.

Techniques of quality management BS 7850: Part 2: 1994 'Total quality management – guidelines for quality'.
Helpful descriptions of simple analysis tools for decision making and problem solving.

Economics of quality BS 6143: Part 1: 1992, 'Process cost models' and Part 2: 1990 [1997] 'Prevention, appraisal and failure model'
The models offered are often more relevant in concept than application.

Preparing specifications BS 7373: 1998.
A standard it would be beneficial to make more use of.

Risk management BS 6079-3: 2000.
See above.

Sampling procedures BS 6001 series (sampling by attributes) and BS 6002 series (sampling by variables).
These describe classic procedures for quality control in mass production. The concepts can be more relevant than the detail except when long production runs are required.

Writing quality manuals (ISO 10013: 1995) and plans (ISO 10005: 1995).
Both documents are limited to guidance and neither is specially useful. Clients often require plans to conform to ISO 10005.

Standards specific to a part of industry ('sector standards')
Design management BS 7000 – in particular Part 4:1996 for design in construction.
The standard complements ISO 9001: 1994. It deals with business and financial management and gives more detail about design itself than is in ISO 9001. In particular there is a useful section on the contents of a design brief and useful annexes on validation, verification and technical information.

Laboratories ISO 17025: 2000.
This standard is compatible with ISO 9001: 1994 and is expected to be revised to be compatible with the 2000 edition. It represents the additions to ISO 9001 needed for a laboratory to meet international standards of accreditation for its testing.

Nuclear installations BS 5882: 1996.
The standard sets stringent requirements for conformity of methods as well as of product and to system. It envisages subcontractors complying with its full requirements and makes heavy demands for traceability.

Project management BS 6079: Part 1: 1996.
This is an approach to managing projects, originally independent of ISO 9001, coming from a mechanical and process engineering viewpoint. It highlights critical path planning, cash flow and achievement. It has much to offer but lacks appreciation of ISO 9001 and of the role of environmental, health and safety systems.

The lists above are not exhaustive – the bibliography in ISO 9001 adds a few standards that could be relevant to other industries but which impinge little if at all on construction.

To make sense of the lists it helps to understand something about standards writing and adoption.

- Standards may start as self-interested industry, public or consumer initiatives to rationalize the market either within a country or internationally. If an initiative finds favour it may be used as the basis of a national or international standard. So at any one time there may be several competing initiatives struggling to establish the interests of their promoters. Where these interests have some broad alignment with the interests of all the parties a standard will emerge. Similarly, national standards are often used to promote a national view as input to an international standard. The result is a mixture of standards, some of which can be viewed as stages in the development of overall consensus, some of which are left behind as consensus develops.
- Generic standards such as the systems standards always face opposition from those who hold that their sector is unique and requires special treatment. This often results in a sector standard. It may offer little over the generic except the use of language familiar to those in the sector concerned and a consistent interpretation of difficult

clauses. The real value is in respected figures within the sector thinking out how to apply the generic standard and so building ownership of the ideas.
- There is an enormous volume of standards writing going on. One result is that standards get written and advance beyond others to which they relate in a random way. New ideas are fed in as they occur and the relationships between standards often depend on the writing programme as much as on the logic of the texts. Over time this will settle down and a long period of stability may ensue: some of the standards made in the 1940s for steel grades are still being used.

The intelligent standards user, therefore, surveys the field and looks to what may emerge as universally acceptable in the 3- to 5-year period. In this way systems can be built to last a substantial time. That's a necessity. Work can only go forward efficiently within a well-understood framework of management.

8.4 Third-party registration and sector schemes

From the 1980s onward the idea has been promoted that management systems can be assessed by independent bodies so that satisfactory systems can be certified and registered. This, it is claimed, gives assurance to would-be customers that the organization is well managed and so avoids or at least reduces the need for costly enquiries about the capability of organizations prior to placing work with them. Early on, it became clear that the probity of firms offering independent assessment and registration would itself have to be assured. This is achieved by accreditation of the assessing firms under national schemes – for example, the UKAS scheme in the UK. But what is assessed remains limited to establishing that the management system is written to meet a standard such as ISO 9001 for quality and that, on the basis of a relatively small sample of work in hand, it is effective. The system is judged effective, moreover, if the declared standards for the work are being met. The fact that the declared standards may be defective, unsuitable or uneconomically high does not affect the certification and registration.

In short, third-party certification and registration of a management system means only that the organization has in place a system that can keep management informed of whether the product is like what it is wished to offer.

A much stronger certification is achieved under the BSI kitemark scheme. In this scheme not only the management system but conformity to a national or international product standard is assessed and regularly monitored by inspection of relevant quality control records.

Between these two extremes lie a number of certification initiatives giving varying degrees of assurance to customers and of more or less usefulness to the organizations that register under them. All the schemes, including the management and kitemark schemes outlined above, succeed only when there are benefits to both organizations and their customers. The principal benefits to customers are improved consistency of product, the potential for the certifier to withdraw registration in the light of serious complaint and sometimes redress through the registering body when an individual firm defaults. The principal benefit to supplying organizations is an improved market, may be in one or several ways – exclusion of unregistered firms, improved confidence, particularly in a new or suspect product, more consistent margins assured by participating firms all meeting a common standard, inclusion in tender lists because customers require registration, risk sharing when complaints are received. Some schemes originate from trade associations, others from powerful customers, a few – mostly concerned with safety – from regulating bodies.

The intermediate schemes of most concern in construction are as follows.

8.4.1 Agrément Board schemes

These offer limited product certification combined with, often but not always, requirements for a quality management system. The schemes suit products such as cavity wall infill or specialist mortars where the method of use is key to the success of the product. In essence, some tests of the end product are agreed with the manufacturer and the Board uses a sample product in accordance with the instructions, tests it to the agreed tests and certifies that the claimed results are achieved e.g. cavity walling material is installed in a cavity and its thermal transmittance is measured to demonstrate that, used as the manufacturer instructs, the product achieves the transmittance claimed. A weakness of the scheme can be that the tests may be few and not very often. A strength is to identify that, satisfactorily used, the product produces the desired effect. In other words, the installation instructions are independently validated.

8.4.2 CE marks

These stem from requirements to assert compliance with the 'essential requirements' of European directives. All the requirements relate to safety and the level of assurance provided depends on what level is acceptable under the directive. Some schemes give a high level of assurance, based on frequent independent inspection, others are no more than a declaration by the supplying organization.

Probably the mark of most importance to the concrete industry is that relating to compliance with the machinery directive. Compliance with this directive is established by the designer/manufacturer following the steps below:

- Make an assessment of the risks to which operators/users and bystanders may be subject.
- Arrange the design to reduce the risks so far as practicable.
- Write a user manual.
- Compile the technical construction file, assembling the design data (drawings, calculations, test requirements etc.) that can be used to demonstrate the machinery meets the directive.
- Draw up a declaration of conformity in prescribed terms.
- Affix the CE conformity marking to the machine.
- If necessary for the safety of personnel operating the machine, attach pictograms in appropriate positions on the machine.

For most machines all this can be done by the designer/manufacturer or importer. But for a limited list of machines thought specially dangerous (e.g. most woodworking machinery; locomotives and some other machines for working underground; lifts for people where a fall of more than 3 m could occur) the declaration and marking must be by an independent approved body.

Because safety and ability to do the designed task are often interrelated for machinery, the mark gives some assurance that the machine is fit for some of its purposes. But it's very important to realize that this is incidental: if a function has no implications for the safety requirements of the directive it won't affect the mark. Much machinery has to comply with other national regulations as well. Again, it's important to realize that the CE mark gives no guarantee that these have been met.

The Construction Products directive is also relevant. Construction products are those that are to form a permanent part of a structure, so concrete is included. Materials have to comply with the fundamental regulations during an economically relevant lifetime, subject to suitable maintenance. The fundamental regulations deal with six aspects:

- Mechanical strength and stability.
- Fire safety.
- Hygiene, health and environment.
- Safety of use.
- Sound nuisance.
- Energy savings and heat retention.

These of course have to be interpreted for each material in much detail. What is easier to grasp is the similarity with building regulations and hence the expectation that materials acceptable under such regulations – generally those acceptable under National or European standards – will be acceptable under the directive.

8.4.3 Sector schemes

While some of these schemes originated from trade associations and others from clients, in the UK they have all benefited from one major client, the Highways Agency, specifying that some products must come from and sometimes be installed by registered firms. Moreover, the chequered origins of some schemes where different certifiers had different requirements are giving way to schedules of requirements administered by national bodies such as UKAS. The success of the schemes rests too on the wisdom of those who put together two early schemes, both coming from trade association bases. These are the CARES scheme for reinforcement supply and the QSRMC scheme for supply of ready-mixed concrete.

It is useful, first, to look at what sorts of requirements are commonly made. They can be categorized as relating to:

- Documentation – definition, verification and review of orders, records of items supplied including traceability to raw materials, part manufacturer's batch etc., records of work done – methods, personnel, inspection etc.
- Definition of end product requirements.
- Competence – qualification and testing of people, sometimes involving regular proving of professional development and/or continuing competence, but particularly related to skill of operatives and first-line supervision.
- Use of equipment meeting defined standards.

- Defined work methods and standards.
- Complaints procedures and remedies.

Different schemes emphasize different aspects, most have some relationship to safety, all act to produce more competent products more consistently. Set out below are the key aspects addressed by the most important schemes in the concrete field.

1 *Supply and installation of post-tensioning systems in concrete structures (CARES).*
- Classifies structures into 'highway' and 'non-highway'.
- Quality system to ISO 9001 audited regularly by CARES, with a compliance element in the auditing.
- Documented quality plans for each project.
- Use of defined materials.
- Traceability of all materials.
- Documented procedure for assessing grout suitability.
- Documented procedures for each installation activity.
- Training and competence requirements for staff and operatives.

2 *Readymixed concrete (QSRMC, BSI)*
- Quality system to ISO 9001 audited regularly by QSRMC/BSI.
- Detailed technical procedures covering all phases from enquiry to delivery. (BSI is less prescriptive than QSRMC).
- High standards of testing (QSRMC/BSI).
- Training and competence requirements for staff and operatives. (Less emphasis in the BSI scheme.)
- Complaints investigation by certifier (QSRMC).

3 *Reinforcing steel (CARES)*
- Quality system to ISO 9001 audited regularly by CARES.
- Surveillance monitoring by CARES.
- Stringent inspection of steel sources and use with traceability requirements aided by rolled in marks on bars.
- Materials test requirements.

The three schemes above are mandatory in current UK highway specification. There is another scheme, that for **precast concrete masonry** operated by CARES, which is not. It is an old scheme. Update is in progress to extend the scope to all precast work and to include additional requirements. When that has happened it's to be hoped it too will be called up in specifications. The principal requirements of the present scheme are:

- Quality system to ISO 9001 audited regularly by CARES, with a compliance element in the auditing.
- Requirement for statistically based product compliance regime to be defined and to include use of CARES approved reinforcement supply.

The continuing success of these schemes will depend on their ability to tackle non-compliance, to be international – CARES with its registration of steel mills worldwide currently leads the field – and to offer continuing improvement within each scheme. This is a challenge for QSRMC, to some extent addressed by the introduction of designated mixes.

8.5 Self-certification and quality control

'Self-certification' is a term used when a construction contractor is responsible for certifying the works he has constructed are correct. It is the responsibility of the contractor's board and project manager as much as is building the works. Self-certifying contracts are often to design and construct the works, and that is what is assumed below. But it is possible for a construct-only contract to be self-certifying and similar requirements apply, but of course only to the construction phase.

To achieve self-certification requires sufficient verification activity to be satisfied that the works meet the employer's requirements as envisaged in the design put forward by the contractor and accepted by the employer's representative on his behalf. Additionally, intermediate verification is needed to safeguard the risks in allowing work that is wrong to go forward when, in the end, it will prove unacceptable. To achieve this and at the same time meet in-house efficiency requirements, demands careful planning and effective, well-supported execution.

The principal planning tools are:

- *Understanding* of hazards and evaluation of risks. This is done formally for environmental, health and safety matters. It may be done less formally for risks to quality and progress. But however it is done, it is risk appreciation that drives economic quality control.
- *Method statements* defining the processes to be carried out – how the works are to be built, with what equipment, in safety and with due regard for the environment.
- *Quality control programmes* in which all the verification steps are defined from design input through to the point of handover.

Drafting quality control programmes is the responsibility of a small team comprising the relevant senior line manager, the engineering manager, the designer's representative and the quality manager. They can be greatly helped by comment from other senior managers, both within and outside the contractor. The work of drafting is a creative activity, allowing input and gaining commitment from all involved. It is not merely the administrative re-ordering of information already in existence. In particular, there is opportunity in writing quality control programmes to confirm the acceptance criteria both for individual programme items and for each payment activity.

Quality control programmes define, for each line of the programme, what documents will be completed and signed to certify that line. The documents are always one or more of the following:

- A certificate of conformity, e.g. for bridge bearings.
- A test result, e.g. for a pile following its installation.
- A qualification certificate, e.g. for a welder.
- A completed and signed inspection and test plan ('checklist' – see Figure 8.12), e.g. for a concrete pour. Drafting these is best made the responsibility of the relevant section engineers, within the guidelines provided and with the assistance of the quality and engineering managers.

Who carries out each verification and signs on the project manager's behalf that it has been done and the works are correct are matters for decision by the drafting team. Often there will be two decisions – who is to certify and who to witness. These are easy

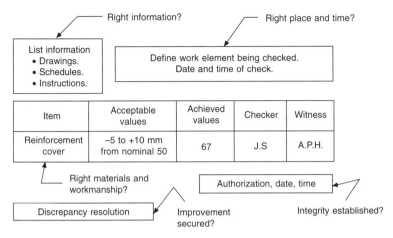

Figure 8.12 Certifying documents: a typical checklist layout.

decisions for all but the last item on the list above. For this item choices are made line by line, based on analysis of risk as shown in Table 8.2.

Table 8.2 Analysis of risk

Category	Description	Examples	Checking level
1	Work critical to structural integrity or safety	Bridge design. Pile butt welds	Independent third-party check
2	Critical operations	Superstructure pour shutter moves	Internal independent check
3	Other work	Normal concrete pours	Supervised checking

Levels of check are subject to review in line with the principles of statistical quality control.

Choices of checker and witness are planned in such a way as to give maximum ownership of the process to the constructing team, aligning the interests of all concerned with the overall interest of right first time construction to give the design service life in an efficient way.

Particular attention is given to the question of *diagnosis and improvement.* This is furthered by selecting inspection and test methods which give useful information about the process and facilitate analyses of discrepancies. Improvement is brought about by reviewing and changing method statements and inspection regimes on the basis of experience: the observational engineering approach epitomised by the phrase *see, judge, act.*

All concerned need to be satisfied that the verification work achieved is genuine so that the certification is well founded. This results in a second tier of work and documentation in support of the first. The elements of this work are:

- Third-party accreditation of, for instance, test houses under an appropriate national scheme, such as the UKAS scheme.
- Internal, second- and third-party quality audit both of processes and of the inspection and testing associated with them.

- Evidence of qualification and professional status of work people and staff.
- Evidence of calibration status of equipment.
- Evidence of the authority and responsibilities of those signing verification documents.
- A provable and convincing chain of documentation linking work, verification inspections and tests to certificates of work completion. Instances – does an inspection sheet record measurements rather than make assertions of compliance? Are signatures related to examples provided and names in the quality plan (or an appended schedule)? Are samples traceable through a chain of custody from being taken to test result?

This second tier of work is primarily the responsibility of the site quality manager, functionally responsible directly to a senior off-site manager or director.

All these steps demand the commitment and goodwill of everyone involved. Their success depends on this. But this is not enough. Self-certifying work can only be successful where expert staff are employed: those concerned must have a high level of technical knowledge backed by substantial experience, giving them insight not only into the immediate but also into the long-term consequences of quality control decisions.

8.6 Details of ISO 9001

8.6.1 General

Much of ISO 9001 is straightforward and, if explanation is required, there is sufficient in the guidance code, ISO 9004 and the consideration of the general text of the standard above. A few matters, however, merit explanation in relation to the concrete industry and the construction industry in general.

8.6.2 System and product review in the management standards

ISO 9001: 2000 requires regular review of the system for its continuing suitability, adequacy and effectiveness by 'top management'. This is called 'management review' and is the subject of section 5.6. It is one of the chief ways in which expression can be given to the principles of leadership, involvement of people and continual improvement. That it includes setting and review of targets is explained in the guidance standard (ISO 9004).

The standard also requires review of requirements related to the product (cl. 7.2.2). This review is required to be made prior to commitment to supply either the original product or any change to it.

These two sorts of review together can provide practical evidence of the operation of the continuous improvement cycle at every level and across all company operations. In construction, they need to be carried out reiteratively with each level of management from directors downwards.

There are similar requirements for review of the environmental, health and safety systems. Management review applies directly to all three systems. For quality, review related to product means contract review. For environment, health and safety it can be taken to mean review of outcomes. These are often negative – incident analyses – but can be positive, e.g. the success of an improvement initiative.

The practicalities of a typical system can be understood from Table 8.3. 'QESH' in the table means the quality, environment, health and safety department – that is, the department responsible for monitoring and maintaining the systems.

Table 8.3 A typical management review

Level	Management review	Product-related review
Company	(6-monthly) QESH meetings with director, including formally reported regular reviews of system documentation.	Discussion of significant matters at management review.
Departments, business units. (3- to 12-month intervals)	Line management meetings with QESH representatives.	Discussion of significant matters at management review. Tender review meetings.
Project (weekly basis)	1 Meetings as described in the project or local unit management plan.	
Project: other actions		2 Initial review of contract against tender documents. 3 Ongoing review of programme, information requests, instructions, etc.

Model agendas for review meetings are provided at each level. These agendas meet and in fact often exceed the requirements detailed in the standard and the ISO 9004 guidance to it. The requirements for management review are given in scn. 5.6 and are to look at inputs to the system:

- audit results, customer feedback, work methods and product conformity, corrective and preventive actions, follow up, changes that might affect the system (e.g. in the market), possible improvements.

It's useful also to review human resources (training, recruitment) and other resources and to look at communication certainly within the organisation and maybe with outsiders – customers and public – too.

Detailed requirements set for contract (or product related) review are set in cl. 7.2.2, namely that:

- Product requirements are defined.
- Contract requirements differing from those previously expressed are resolved.
- The organization has the ability to meet the requirements.

8.6.3 Purchasing, subcontract and materials control (ISO 9001: 2000 scn. 7.4)

In ISO 9001 terms, the aim of the purchasing process is to make sure that what's purchased is satisfactory on delivery. At each stage this aim is expressed in terms of verifying the capability for the goods or service to be satisfactory, in terms of resources, system and expertise. There are also underlying statutory duties: health, safety and environmental requirements have to be met. And there are commercial requirements to give an acceptable return to shareholders.

So in the real system of a commercial company 'satisfactory' takes a broad meaning including not only quality of goods but timely and reliable delivery, efficient dealing with problems, meeting health, safety and environmental requirements and acceptable price and business terms.

In outline, purchasing to meet all these requirements is simple but reiterative, involving appraisal of suppliers at five stages:

1 General. Find possible suppliers and determine the capability of each.
2 Particular. Review offers to select supplier.
3 Prior to start. Verify arrangements, quality of product, programme.
4 In course of supply. Review to maintain performance.
5 On completion. Review to provide feedback for the future.

All these stages are obligatory under health and safety legislation to establish that health and safety arrangements can be and are met. Environmental legislation makes the main contractor so liable for the actions of suppliers that all stages are desirable. The links between doing work as planned and accident, incident or nonconformity are so strong that it is undesirable to omit any stages in the appraisal of quality.

At each stage there are only three questions:

- What is needed to support the performance of the supplier?
- Allowing for whatever is needed is the deal still commercially satisfactory?
- If it isn't what are the options?

In practice, in an organization with a few regular suppliers, such as most design firms, the system to give effect to these requirements can be simple too. Operation of the system is often lax, however, because the risks of leaving out stages are seen as low compared with other matters such as delivery date or profit. Also, appraisal is an objective process, often felt to be at odds with the personal relationships that develop between supplier personnel and organizations. The feeling 'they have done well for us in the past' often outweighs the desire for objective assurance based on present capability.

In a large construction contractor with a large and often changing base of suppliers the system becomes complicated. In particular it has to cater for the appraisals getting out of phase – particular appraisal may precede or take the place of general appraisal – and for a wide variety of staff being involved, sometimes with site staff substituting for buyers or vice versa. Moreover, the work load of full appraisal of every supplier would be inordinate for the benefits gained. What has to be introduced are measures to control this. There are two:

- Development of preferred suppliers, allowing appraisals 1 and 5 to be time rather than order based and limiting what is needed at the other stages.
- Risk assessment to determine extent of appraisal at each stage. The time to do this is when writing the project management plan. The methodology is that set out in the general consideration of ISO 9001 above.

8.6.4 Nonconformity and improvement

Definition of nonconformities is always an issue. There are three common problems:

1 *Does a nonconformity arise before work is finished, or only at the point of contractual handover?* Management has choice here. Of course once work is completed and handed over to the customer any discrepancy from specification is a nonconformity. But internally the choice can be made that the occurrence is so insignificant and rare as not to merit investigation or, alternatively, is of sufficient significance or occurs sufficiently often to merit the costs of recording and bringing about improvement. Management can only know the rarity of occurrences, however, if all potential nonconformities are recorded initially. Beyond that, the point of identifying something as a nonconformity is to facilitate improvement: be guided by that aim.

2 *What is actually acceptable/unacceptable?* For some aspects of concrete this is well understood – there are seldom arguments about what 'strong enough' or 'of the right dimensions' mean. But arguments about appearance are common and so are arguments about whether some of the less common test results indicate conformity or nonconformity. Understanding about measurement and sampling uncertainties (Figure 8.13) helps, so does understanding that the best can never be the norm.

3 *Is the classification of something as 'nonconforming' counterproductive?* It may often be so – no-one likes being told they are wrong and 'wrong' and 'nonconforming' are often linked together. It may well be better to use a less judgemental word such as 'problem'. But any word becomes judgemental if that is the attitude of those involved. Good leadership fosters the better approach that when there is a problem we all share in it, in solving it and in preventing recurrence.

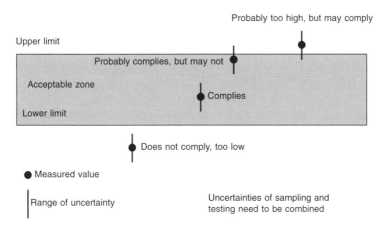

Figure 8.13 Quality control: uncertainty of measurement.

All nonconformities have to be handled as set out in Figure 8.14. There are no other possibilities.

The features of the figure that come from quality management principles are those related to prevention.

The idea that it isn't enough just to put something right, that it is essential to seek a way of preventing the same thing going wrong again is a key insight of quality management.

Quality concepts

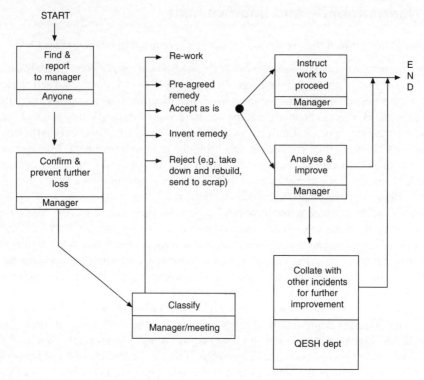

Figure 8.14 Handling nonconformities.

What often acts against the idea being given effect is guilt: *It's come out wrong. It must be because we've done it wrong, let's try the same again, doing it 'properly' this time.* Much time is wasted because no-one has the confidence it was done right first time so no-one has the confidence to follow a path of logical fault finding and improvement.

The other prevention-related feature is collation: looking at other similar problems, correlating one problem with another, looking at the aggregate of problems as well as individual problems as they arise. This is a powerful process, reducing guilt, offering new possibilities for prime causes, avoiding remedies already proved false elsewhere. Failure to consider problem interactions, to collate information, on the other hand, can be negligent.

Figure 8.14 points to the opportunity to have planned solutions for likely nonconformities. This is sometimes thought to be tempting providence. Yet in both environmental and health and safety spheres it is plainly negligent not to have incident procedures both in place and rehearsed. There is, it is pleasing to say, a growing literature of plans for the technical handling of nonconformities. Examples are:

- The CIRIA report *Action in the case of nonconformity of concrete structures* (1997).
- The Dutch system of financial penalties for nonconformity in blacktop.
- Remedial requirements written into welding standards.

Such documents are the logical outcome of applying risk management principles. They are well worth adding to on a project by project basis.

Improvement can be pursued independent of fault or nonconformity. Indeed it is essential that it is because what was right yesterday will become not good enough tomorrow as markets change, client expectations change, technical changes alter risks. Only by planning for continuous improvement can a firm's market share be maintained and its success be steadily enhanced.

Good management therefore requires the provision of mechanisms for continuous improvement in addition to those that are driven by the current run of nonconformities. Quality systems address this need by making requirements for:

- Data collection, e.g. of life cycle costs, satisfaction of customers, employees, the public, regulatory bodies, process costs, benchmarking.
- Defining measuring and monitoring processes.
- Planning.
- Providing techniques for bringing about improvement, e.g. brainstorming, paired indicators, Pareto diagrams. These are set out in various standard texts such as Oakland's *Total Quality Management.*
- Creative use of audit.

Beyond these mechanisms, continuous improvement depends on establishing a climate of opinion within which change is treated positively. This too is addressed within quality standards, through requirements for management commitment, communication etc.

8.6.5 Audit and review

Organizations and auditors are likely to be involved in three different kinds of audit.

- **Appraisal** By a 'third party' – that is, someone other than the firm or its customer – to decide whether the firm can be certified as meeting prescribed quality standards.
- **External** By a 'second party' – that is, a customer, potential customer or customer's representative – to decide whether the firm is well enough organized to be able to meet the quality standards required.
- **Internal** By a 'first party' – that is, a member of the same firm or group to help the firm meet and improve on its own quality standards.

In each of the above definitions the words 'quality standards' occur. Commonly there will be three sorts of standard to consider:

1 **Application** and their backing standards: Shape, finish, level of excellence, meeting test requirements. There are lots of such standards – for bridges, etc. Sometimes the standard is expressed as a specification particular to the job or class of work. Sometimes it is very simply 'one that works'.
2 **Regulatory** standards   Defining the safety, health, environmental and other requirements acceptable to society in general. These may include legal and insurance requirements, voluntary, national or trade codes, standards set by the firm itself to establish its position as a **good** firm.
3 The **system** standard   Usually ISO 9001, a single standard describing how an organization can best manage its affairs so that the other two sets of standards can be met.

Traditional financial auditing is 'third party' and the certification is to assure shareholders and the public that the financial affairs of the firm concerned are truly stated (meet the product regulatory and system standards). Modern financial auditing can only achieve this by looking first, at the **system** for producing financial statements, second, at the **principal transactions** and third, at a **sample of lesser ones**. While it remains 'product' oriented it embraces system assessment as a means of checking that sample transactions are typical of the whole and that the system ensures regulatory requirements are met.

Quality auditing in general concentrates on looking at the **system of management.** Meeting regulatory requirements and the correctness of individual product are assessed incidentally to the auditing process. However, it is well known that being able to do well is different from actually doing well. Wise quality auditors make sure that they do not approve a system under which a bad product is being produced, or in which bad regulatory practice is tolerated.

Requirements for auditor qualification and training and for the conduct of audits are given in ISO 19011. In summary:

- Auditors are trained.
- Audits are prepared for by comparing relevant data with relevant standards and writing a list of questions.
- Audits start and end with meeting the senior people involved. In between, work is observed, documents are inspected and people explain what they are doing.
- The auditor compares what he finds with what was expected and makes his comments as he goes along.

Essentially the same procedure is applied to quality, environmental, health and safety audits. Audit procedures need to be backed by technical guidance about particular sorts of audit, e.g. auditing design firms, readymix companies. Where auditors lack such guidance time will be wasted either because of their lack of technical knowledge or because they will lack the self-knowledge that comes from testing experience against teaching.

The auditor will report differences between promise and practice as he finds them, often issuing an improvement request there and then. He will report his findings in total at a closing meeting to a senior representative of the firm. His written report should contain nothing not already stated verbally. External auditors give no more than **findings** of fact about the system compared with the standard and if appropriate a **conclusion** such as 'This works could supply satisfactorily'. Auditors can and usually will make **recommendations** to their own firm: about what steps can be taken to ensure a would-be supplier's product will be satisfactory; about how a problem perceived on internal audit could be dealt with. These are statements about what the auditor would do if he were in charge: useful because they ensure attention is directed to finding a way forward, but leaving the line manager free to come to his own conclusions. Audit reports always include a note of how recommendations are to be cleared; that is how and by when the management concerned are to let the auditor and the QESH department know what actions are in hand.

Audits cost money and are only of value if action follows from them. The first step – common sense as well as required under the standards – is that they are carried out to a programme. Sometimes this is set in the systems department, but it is better set and reviewed by line managers, with the systems department keeping a check on its competence. The process of achieving action is carried out by the line managers concerned and is called **review**. Managers consider all the data – from audits and other sources – decide on actions to take, make arrangements to implement them, establish a means of judging their effectiveness and a means of making further changes if these are desirable. This review audit by audit is best supplemented by review of the run of audits at divisional and company levels. An overall view of audit usefulness and a framework for detailed planning of future audits can then be formulated and kept under review.

8.7 Laboratory management

8.7.1 Introduction

Quality systems for testing and calibration laboratories are defined in ISO 17025: 2000, *General requirements for the competence of testing and calibration laboratories*. In the UK and across Europe this standard is replacing previous national regulations. In the UK it replaces NAMAS publication M10 *General criteria of competence for calibration and testing laboratories*.

The standard contains 'all the requirements that testing and calibration laboratories have to meet if they wish to demonstrate that they operate a quality system, are technically competent, and are able to generate technically valid results'. It supplements ISO 9001 (at present the 1994 edition) with specific process-related requirements. It is very stringent. It is costly to implement in full. It is directed towards the provision of a standard of testing and reporting that can be depended upon in a criminal court. The high cost rests, however, not on the system of management, but on the levels of calibration, recording, reporting and security required.

Some work requires full accreditation – cube testing is the most common example – but any testing that may be required to support a legal action is best accredited. For much routine testing, however, what is needed is reliability within known limits at a moderate level of proof. If this testing points to errors it can be repeated or other tests can be carried out to verify the situation. There is little point in establishing very high levels of accuracy and proof for testing when the sampling regimes to provide test specimens are poor or are subject to much larger errors than the testing itself.

The ISO 17025 management system needs to be applied to all sampling and testing work. Individual requirements – e.g. for level of calibration – can be relaxed to suit the situation, but there is no excuse for not applying the framework of good management and good sampling and testing practice.

8.7.2 The ISO 17025 framework

The framework requires positive management of the following:

- **Overall system**   Policy, documentation, communication, corrective and preventive actions, improvement.
- **Environment**   Temperature, wind, shelter, access, security, work space, etc.
- **Samples**   Methodology, identification, quality, chain of custody, care to preserve condition, sub-sampling, disposal.
- **Calibration**   The quantity of uncertainty associated with any equipment used and so how dependable any result could be. Calibration is a two-stage process: full formal calibration at regular intervals and checks when used that the calibration is still valid.
- **Testing**   Procedures, work sheets, identification of equipment and personnel, quality control.

- **Personnel** Training, qualification, continuing assessment.
- **Reporting results** Written chain of evidence, factual statements, reporting differences from sampling or test requirements, uncertainty of measurement, confidentiality, review, authority.
- **Opinions** These are to be clearly identified and the basis on which they are given is to be declared.

This is a list of thirty-nine items. Every piece of measuring or testing with which you are involved requires consideration of each one so that you can know how much to depend on the answers you have.

8.7.3 Management of concrete sampling and testing

The QSRMC scheme sets high standards for testing concrete in laboratories, approximating to the requirements of UKAS accredited testing to ISO 17025 and similar standards are approached for field testing by UK readymix companies. Customer – that is, generally construction contractor – testing has traditionally been less rigorous on-site, though many major contractors require off-site laboratory testing to be carried out under UKAS accreditation. The UK Highways Agency and some railway contracts now require, for major work at least, that slump and cube making on site and almost all off-site laboratory testing is accredited.

The site requirement is difficult to meet economically unless a major site laboratory and experienced materials engineer is required full time anyway. What is really needed is a simplified scheme, overseen by the relevant national body (e.g. UKAS), that covers all site testing and more importantly sampling, enabling local staff to be readily and quickly trained and qualified on a job-by-job basis. This scheme could be driven by risk rather than an artificial requirement to meet the level of proof required in a criminal court and all parties could subscribe to and cooperate in it.

It's interesting to look at where a risk-based system might lead. Currently around half of cube strength problems are attributed to sampling/specimen error and about a quarter are never satisfactorily explained. Moreover it's rare for concrete to be condemned unless it is patently defective from an obvious cause. My conclusion is that a system that concentrated more on sampling and initial inspection and less on cube strength testing to high accuracies would give better value for money.

Of course, a sufficiently high level of testing would also be required to resist fraud. This testing, though, requires a specific plan, different from quality control testing. What is needed is related to contractual arrangements, the strength of self-regulating schemes such as the QSRMC and local circumstances.

References

Argenti, J. (1989) *Practical Corporate Planning*, Unwin, London.
BSI (1994) BS 7850: Part 2: Total quality management – guidelines for quality improvement. British Standards Institution, London.
BSI (2000) BS 6079-3: Project management – Part 3: Guide to the management of business related project risk. British Standards Institution, London.
Oakland, John S. (2003) *Total Quality Management*, 3rd edn. Butterworth-Heinemann Oxford.

9

Quality control

Lindon Sear

9.1 Introduction

The variations in a manufacturing process could have serious technical, quality and commercial implications to a company if not monitored and controlled in some manner. Statistical analysis of data using various formulae prior to the widespread introduction of the computer was practically impossible. To calculate the mean and standard deviation from a dataset using longhand was slow and impractical. Simplified techniques were required in order to control manufacturing processes.

Controlling the quality of readymixed concrete is similar to any process control system. The nature of the raw materials used in the manufacture and the large number of factors, which affect the strength of readymixed concrete, makes it a highly variable product. The coefficient of variation is typically between 10 per cent and 20 per cent. A further complication is the variety of the concrete mixes used in the construction process. These mixes vary in cement content, cement type, strength required, aggregate size and type, workability and from the effects of the addition of performance-enhancing chemicals. To make easier the task of controlling such mixes requires simplification of the data. A number of systems exist which can achieve this aim as follows.

One 'control chart' approach was developed by W.A. Shewhart in 1924 (Mendenhall and Sincich, 1988). Shewhart charts are used by some suppliers in the United Kingdom and are common in the United States of America as a control system for strength mixes. They can be difficult to interpret when changes in more than one property of the concrete has occurred, e.g. simultaneous changes in mean strength and standard deviation (SD) of the concrete. Such objections can be overcome with complex analysis of the data.

Another control chart approach is the Cusum system of quality control. This was developed by ICI in the 1960s as a wide-ranging control system (Woodward and Goldsmith,

1963). In the 1970s, RMC (Brown, 1984) developed a version specifically for controlling the production of readymixed concrete. It is the most prevalent technique used by readymixed companies in the UK.

Though we shall only cover the two systems above, other control systems exist. One such other system developed by the British Readymixed Concrete Association (the BRMCA is now part of the QPA; see Dewar and Anderson, 1988; Barber and Sym, 1983)), is based on the analysis of the positive and negative deviations from the target mean strength. When several identical signs occur concurrently the mean of the last 10 results is calculated and the required adjustments to the target mean strength made. More complex Cusum systems can be created to control multiple variables (Sear). With the ever-increasing power of spreadsheet software, direct analysis using fundamental principles is practicable.

9.2 Control charts

Control charts are constructed from tests on samples of the product. The product could be anything ranging from light bulbs to concrete. Graphs are normally used to present this data, as we all know 'a picture tells a thousand words'. By plotting the product's 'quality variable', i.e. the test result values, over a period of time a visual representation of changing quality with time is provided. The quality variable could be the longevity or the current consumption of a light bulb or the strength of a concrete cube. To be successful a control chart should be able to distinguish between:

- The random variation within a process, e.g. the normal variability of the sampling procedures. In concrete terms this would be effects of the concrete technician sampling, the testing machine, the curing regimes, etc.
- An assignable variation, e.g. a change in mean strength or workability of the concrete mixes. These may result from changes in cement performance, aggregate quality, weather conditions, etc.

Plotting the quality variable, the cube strength for our purposes, against time may give some information about a process. Figure 9.1 uses data set 1 from Table 9.1, which is a set of 28-day cube strengths.

Perhaps these data indicate a trend in mean strength, as there is a gradual increase in strength between 5 Jan and 11 Jan? Is this apparent trend due to random variation or does it indicate a significant change in mean strength?

Now look at Figure 9.2. What is different between the data sets 1 and 2 as used in Figures 9.1 and 9.2?

Figures 9.1 and 9.2 are the same data set except 8 MPa has been added to strength results between 4 Jan and 9 Jan inclusive. This large change in strength would be classed as an assignable variation. This presumes we can differentiate the effect from the random variation. The similarity between Figures 9.1 and 9.2 demonstrates one of the difficulties of statistics and the interpretation of simple control charts, although the change in strength is large.

We shall look at ways of constructing and operating control systems that are able to resolve differences between random and assignable variation, the first system being Shewhart charts.

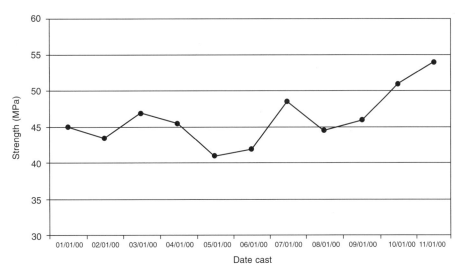

Figure 9.1 A plot of strength with time provides some information.

Table 9.1 Dataset used in Figures 9.1 to 9.4

	Dataset 1 as used in Figure 9.1	Dataset 2 as used in Figures 9.2 and 9.3	Dataset 2 as used in Figure 9.4
Date cast	Strength @ 28 days MPa	Strength @ 28 days MPa	Range of pairs
01-Jan	45.0	45.0	
02-Jan	43.5	43.5	1.5
03-Jan	47.0	47.0	3.5
04-Jan	45.5	53.5	6.5
05-Jan	41.0	49.0	4.5
06-Jan	42.0	50.0	1.0
07-Jan	48.5	56.5	6.5
08-Jan	44.5	52.5	4.0
09-Jan	46.0	54.0	1.5
10-Jan	51.0	51.0	3.0
11-Jan	54.0	54.0	3.0
	Average	50.55	3.5
	Average of last 5 results	53.60	

9.3 Shewhart charts

9.3.1 Monitoring the mean strength

As shown, it can be difficult to discern significant changes in a variable simply by looking at a graph. We need to construct some system that indicates when a statistically significant change in the process has occurred, a control limit. Within any production process, we have target values for the product. For concrete, we have a Target Mean Strength (TMS) and Target Standard Deviation (TSD). For concrete mixes, the TMS and TSD are related to the characteristic strength by the 'margin', normally $2 \times$ TSD. In order to produce concrete with a failure rate of 5 per cent as required by many standards, e.g. BS 5328,

9/4 Quality control

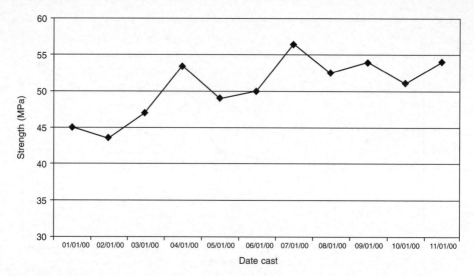

Figure 9.2 What does this data set tell us about our concrete?

BS 8110, etc. the mean strength and standard deviation of the production concrete must be monitored to ensure compliance with the failure rate.

From the statistics of normal distributions, we know that a proportion of our results must fall within a 'margin' around the mean value. We have learnt that 95 per cent of the results must fall within ±1.96 standard deviations (two-tailed test). There is a very low probability of any results falling outside ±3.0 standard deviations (0.3 per cent) from the mean strength. Therefore, a simple control can be constructed with lines indicating these probability levels as in Figure 9.3. These are called the Upper and Lower Control Lines or Action Lines.

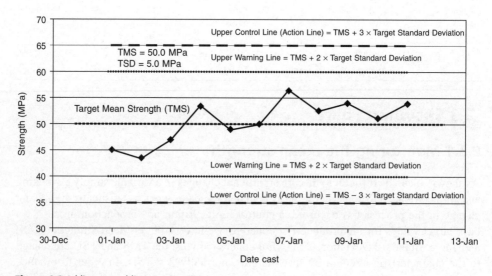

Figure 9.3 Adding control limits at 3 × TSD.

An additional pair of control lines is often used called the Warning Lines. The probability of a single result falling outside of ±2.0 standard deviations is 4.57 per cent. The probability of two results falling outside these lines is only 0.21 per cent (two-tailed) and only 0.05 per cent for a single-tailed test, i.e. in the same direction.

From Figure 9.3, it is now clear that no significant change in mean strength has occurred as all the results fall within the expected band of strengths. As no result falls outside of the Upper and Lower Warning line there is a very low probability of a change in mean strength. However, there may be some indication that the mean strength was somewhat low of the mean for results between 1 to 3 January and the results after 7 January imply a trend of improving mean strength.

The control and warning lines for a Shewhart chart rely heavily on the TSD. Therefore, we must monitor the standard deviation of the dataset to ensure the TSD is a reliable estimate of the population SD. For this we construct another chart.

9.3.2 Monitoring the standard deviation

To compute the standard deviation from any dataset is a complex calculation. However, from the 'Method of Ranges' we know that the range of a group of results is directly proportional to the standard deviation of the dataset. Normally the range of consecutive pairs of results is used, though larger groupings may be advantageous with the Shewhart system. From the 'Method of Ranges' the Target Range (TR) and Standard Deviation (TSD) are related as follows:

$$TR = TSD \times 1.128$$

To construct a Shewhart chart for the range we need to know the standard deviation of the range (SD_R). This is simply related as follows:

$$SD_R = K_R \times TSD \text{ where } K_R = 0.853 \text{ as in Table 9.2}$$

K_R is based on the distribution of ranges, which is a lognormal (also known as a log Gamma) function related to the F test. Even though the distribution of the range is lognormal, the standard deviation of the range is normally distributed, hence why a single constant, K_R, is applicable. Therefore, we can construct our chart with Upper Control Line (UCL), Lower Control Line (LCL) and Upper Warning Line (UWL), Lower Warning Line (LWL) from the following calculation:

$$UCL \ \& \ LCL = TR \pm 3SD_R = TR \pm 3 \times 0.853 \times TSD$$

thus

$$UCL = 1.128 \times TSD + 3 \times 0.853 \times TSD = 3.687 \times TSD$$

and

$$LCL = 1.128 \times TSD - 3 \times 0.853 \times TSD = -1.431 \times TSD$$

Note: LCL cannot be less than zero as the range of pairs is an absolute value, i.e. always positive. LCL cannot be plotted for ranges of pairs.

Similarly, UWL and LWL can be calculated, though, as before, LWL cannot be plotted:

$$UWL \ \& \ LWL = TR \pm 2SD_R = TR \pm 2 \times 0.853 \times TSD$$

thus

$$\text{UWL} = 1.128 \times \text{TSD} + 2 \times 0.853 \times \text{TSD} = 2.834 \times \text{TSD}$$

and

$$\text{LWL} = 1.128 \times \text{TSD} - 2 \times 0.853 \times \text{TSD} = -0.578 \times \text{TSD}$$

Note: Again the LWL cannot be plotted.

Figure 9.4 shows the range of pairs chart for the dataset used for Figures 9.2 and 9.3 These data show the actual SD is lower than the TSD with no results greater than either UCL or UWL. However, as only two results exceed the TR line one interpretation would be that the TSD is too high. Clearly, further analysis is required for both mean strength and range to determine whether any trends are indicated in the dataset.

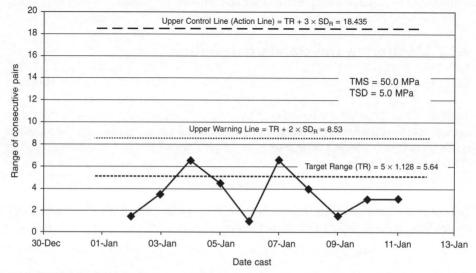

Figure 9.4 Shewhart chart for range.

The lack of lower warning and control lines is a drawback to using the range of pairs. Larger groupings must be used to create LWL and LCL criteria. Other values for K_R have been computed for different group sizes, up to 25, for range. Table 9.2 lists these values.

Table 9.2 Values of K_R for various group sizes for range

Group size	K_R	Group size	K_R	Group size	K_R
2	0.853	11	0.787*	20	0.729*
3	0.888	12	0.778*	21	0.724*
4	0.880	13	0.770*	22	0.720*
5	0.864	14	0.762*	23	0.716*
6	0.848	15	0.755*	24	0.712*
7	0.833*	16	0.749*	25	0.709*
8	0.820*	17	0.743*		
9	0.808*	18	0.738*		
10	0.797*	19	0.733*		

* Indicates an LCL will exist for this group size.

9.3.3 Further analysis for trends

To determine trends within Shewhart charts requires further analysis of these data, known as Runs Analysis. As concrete strengths are assumed to come from a Normal distribution, we can analyse the probability of a number of results being above or below the TMS line. As the normal distribution is symmetrical around the TMS there is a 50 per cent probability of a result[1] ($P_{>TMS} = 0.50$) being above or below this value. In the Table 9.1 dataset for Figures 9.2 and 9.3, there are five consecutive results, which are above or below the TMS of 50 MPa. As each has a probability of 0.50, the combined probability can be calculated from the Multiplicative Law of Probability as:

$$P_{5>TMS} \text{ or } P_{5<TMS} = 0.5 \times 0.5 \times 0.5 \times 0.5 \times 0.5 = 0.03125 \text{ or } 3.125\%$$

There is a 3.125 per cent chance that the five consecutive results in the dataset being on the same side, above or below the TMS value, are due to random variation.

The probability of five consecutive results being above the TMS is the product of the probability of five results being above TMS and the probability of them being below the line. From the Additive Law of Probability this is calculated as:

$$P_{consec.5>TMS} = (0.5)^5 + (0.5)^5 = 2 \times 0.03125 = 0.0625 \text{ or } 6.25\%$$

There is a very low probability, only 6.25 per cent, that these results are due to random variation and it is highly likely that there is an increasing trend in mean strength. This result would suggest either the TMS for the mix can be adjusted or, for a given characteristic strength, the cement content of the mix could be reduced.

Runs Analysis can be broken down into some simple rules of thumb as follows: A significant trend occurs when:

1 Seven or more consecutive results are on the same side of the centre line
2 At least 10 out of 11 consecutive results are on the same side of the centre line
3 At least 12 out of 14 consecutive results are on the same side of the centre line
4 At least 14 out of 17 consecutive results are on the same side of the centre line

9.3.4 Conclusions

Shewhart charts are powerful tools for monitoring a process. However, knowledge of statistics is required in order to determine trends within the data. Visually these charts give little information about the process and can be misleading. Simultaneous combinations of trends can be highly difficult to interpret. A change in mean strength coupled with a change in standard deviation requires some expertise to resolve. Figure 9.5 shows a Shewhart chart on which changes in mean strength and standard deviation have occurred. Visually it is difficult to determine when the change has occurred and no 'action' is indicated in the graph!

---

[1] Using fundamental statistics, any result will not be identical to the TMS when an infinite number of decimal places are used. However, with real data where results are rounded to the nearest 0.5 MPa some results will be either above or below the TMS but never equal to it. Such results are better ignored from the calculations.

9/8 Quality control

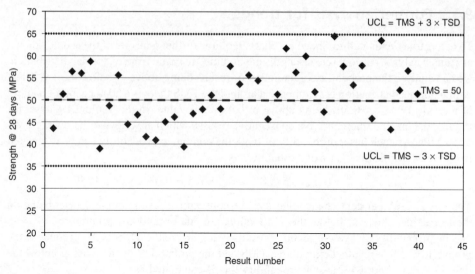

Figure 9.5 A change in mean strength and standard deviation has occurred – where?

The alternative control chart system in common use is the Cusum system of quality control.

9.4 Cusum charts

9.4.1 The history of the Cusum system of quality control

The Cusum technique was developed by Imperial Chemical Industries in the 1960s to control a variety of production systems. The first comprehensive description of the system was published by ICI as Monograph No. 3 'Cumulative Sum Techniques' (Woodward and Goldsmith, 1963) and later was described in British Standard BS 5703. ReadyMixed Concrete Ltd developed the Cusum for application to readymixed concrete and subsequently marketed the system commercially as a manually calculated control system (Brown, 1984). This system was widely adopted from the early 1970s and became the prime method of controlling the strength of readymixed concrete. The Quality Scheme for Ready Mixed Concrete (QSRMC), when formed in 1984, adopted this system as an approved method of quality control (QSRMC, 1991).

The Cusum technique was quickly accepted (Francis, 1997) worldwide due to the simplicity of data entry. Usually the data are entered by Field or Laboratory staff. Experienced staff may have only a working knowledge of the statistics involved and not the complexities of the statistics of Student t testing or F testing, for example. With the Cusum system action points are precisely defined and by cross reference to tables and graphs the degree of changes to cement content, to compensate for changes in strength or standard deviation, are easily found. This makes the Cusum technique ideal for the industry.

9.4.2 Controlling the strength of concrete in practice

The QSRMC and BSI, which are the main third-party quality assurance bodies within the United Kingdom, prescribe the minimum levels of concrete testing within the industry. The minimum requirements for the test rate are for sixteen cube test results per month. There are also restrictions to the range of cement contents used for controlling the quality of the concrete supplied. Where these test rates are not achieved QSRMC requires increased margins on the characteristic strength, (see Table 9.3).

Table 9.3 Margin factors to be adopted – QSRMC regulations

Test rate cube sets per month	Margin factors to be adopted	Minimum design margin (MPa)
16 or more	$2.0 \times SD$	7.0
12 to 15	$2.3 \times SD$	10.0
8 to 11	$2.6 \times SD$	12.0
6 to 7	$3.0 \times SD$	14.0

These test rates were reasonable when a single cement type was in use, i.e. ordinary Portland cement (PC). However, QSRMC's interpretation, when a secondary cement type was being produced, was to require a further 16 test results.

9.4.3 Controlling the mean strength of the concrete

In the RMC commercial system, designed to control a single cement type, the 28-day strengths for the test cubes either are predicted from seven days or accelerated cubes using a series of tables. A description of the prediction system is given below.

From the predicted 28-day strength, the result is corrected back to a control cement content by addition of strength adjustment factor(s). This makes each result appear to the Cusum system as if it is made from the same concrete mix, i.e. the control mix. Comparison is made between this result and the expected value, the Target Mean Strength, and the differences are cumulatively summed.

Table 9.4 gives a Cusum calculation for mean strength. This shows the simplicity of the calculations involved and shows a typical downward trend in the data. When graphs are plotted of the Cusum values, assessment of trends can be made by placing clear plastic overlay masks on the graph. When the graphical plot crosses the action arms, it indicates that a significant change has occurred.

The overlay, a truncated V-shaped mask as shown in Figure 9.6, shows whether a plot displays a significant trend in mean strength. The general mask shape, described in detail in BS 5703: 1980 has an offset of five SD at the 'Lead Point'. The lead point is applied to the last point plotted and the mask slope would be ± 1.0 SD for every sample interval. The masks used for the control of readymixed concrete have somewhat differing values.

For Cusum M, the offset at the lead point, known as the decision interval, is 8.1 multiplied by the current standard deviation and the gradient of the action arms $\pm 1/6$th of the current standard deviation per sample interval.

The response of this control system to changes in actual mean strength is calculated by

Quality control

Table 9.4 Example of a typical Cusum M calculation

Result no.	Predicted 28-day strength (MPa)	Adjustment to control mix cement content (MPa)	Result adjusted to control mix (MPa)	Minus target mean strength of 40 MPa	Cusum of the mean strength Cusum M
1	38.0	+ 3.5	41.5	+ 1.5	+ 1.5
2	43.0	− 5.0	38.0	− 2.0	− 0.5
3	34.5	0.0	34.5	− 5.5	− 6.0
4	36.5	0.0	36.5	− 3.5	− 9.5
5	47.0	− 5.0	42.0	+ 2.0	− 7.5
6	43.0	0.0	43.0	+ 3.0	− 4.5
7	37.5	+ 3.5	41.0	+ 1.0	− 3.5
8	40.0	0.0	40.0	0.0	− 3.5
9	28.0	+ 8.0	36.0	− 4.0	− 7.5
10	33.5	0.0	33.5	− 6.5	− 14.0

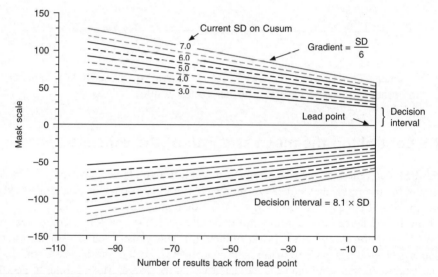

Figure 9.6 Typical Cusum V mask.

resolving the formula for the slope and offset of the V mask. When an action point is detected, the mean slope of the Cusum plot is as follows:

$$\text{Mean slope} = \frac{\frac{\text{SD} \times \text{No.}}{6} + 8.1 \times \text{SD}}{\text{No.}} \quad (9.1)$$

where No. = the number of results back from the lead point

The target mean strength is adjusted by 0.75, an anti-hunting factor, multiplied by the mean slope. The response of this technique is approximately log normal when simulated by computer. Figure 9.7 shows the 95 per cent confidence limits for the run length found from a computer simulation. It shows that to detect relatively large changes in mean strength requires many test results. Jones (1992) suggests that additional tests can be made on the plots using a parabolic overlay mask, which may indicate significant changes

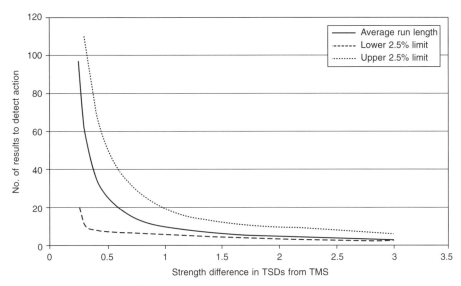

Figure 9.7 Average run length for Cusum M.

more quickly.

9.4.4 Controlling the standard deviation of the concrete

Monitoring the standard deviation of the concrete is again achieved by using the method of ranges. The range of results for various group sizes is directly proportional to the standard deviation. In the Cusum system of quality control the group size is two, i.e. using consecutive pairs of results and hence Range = $1.128 \times$ SD. By comparing the range of the results against the expected range, which is the target standard deviation $\times 1.128$, a Cusum of the range is calculated. Continuing with Table 9.4 data, the calculations are given in Table 9.5. The target standard deviation used is 5.0 MPa and therefore the target range is $5.0 \times 1.128 = 5.625$ or 5.5 MPa to the nearest 0.5 MPa. The Cusum of the ranges is known as Cusum R.

(*Note*: It is customary to work to the nearest 0.5 MPa within the readymixed industry as the repeatability of concrete testing is only within 1.5 MPa even in the ideal conditions of the laboratory.)

Table 9.5 shows increasing negative values, which suggests the actual standard deviation is lower than the target standard deviation. The data are plotted on a graph and, as with the mean strength, a V mask is used to detect a significant change. The design of the V mask is different from Cusum M, the decision interval being $8.5 \times$ target range and the gradient of the action arms being $\pm^1/_{10}$th target range. When the line passes an action arm, a significant change is deemed to have occurred.

Figure 9.8 shows the Cusum M and R plots for Tables 9.4 and 9.5. The action mask is applied when each point is plotted.

Quality control

Table 9.5 Example of Cusum R calculations for controlling the standard deviation

Result corrected to control mix cement content, (MPa)	Range of consecutive pairs from data set, (MPa)	Target SD = 5.0 MPa. Target Range = 1.128 × 5.0 = 5.5 MPa Actual range minus target range of 5.5 (MPa)	Cusum of range of pairs or Cusum R
41.5	n/a	n/a	n/a
38.0	3.5	− 2.0	− 2.0
34.5	3.5	− 2.0	− 4.0
36.5	2.0	− 3.5	− 7.5
42.0	5.5	0.0	− 7.5
43.0	1.0	− 4.5	− 12.0
41.0	2.0	− 3.5	− 15.5
40.0	1.0	− 4.5	− 20.0
36.0	4.0	− 1.5	− 21.5
33.5	2.5	− 3.0	− 24.5

9.4.5 Monitoring the accuracy of the predicted 28-day strength from the early test results

The standard time for assessing the strength of concrete is at 28 days from the date of casting. Clearly to wait for such a period before detecting changes in the quality of the product could have disastrous consequential effects to the integrity of a concrete structure. Therefore, statistical analysis of earlier test results than 28 days is normal practice.

Normally 7-day test results or accelerated tests are entered. There has been limited success with accelerated systems as they normally entail heating the concrete to temperatures of 35°C, 55°C or 82°C (Grant, 1977) in the British Standard method (BSI, 1983) or higher (Lapinas, 1975). Such heating has been found to have quite variable effects, dependent on the pre- and post-curing given (Sweeting, 1977; Dhir and Gilhespie, 1983; ERMCO, 1983), on the strength of the concrete, which is not seen in the actual 28 day strengths. Day (1995) suggests that early age testing at 24 hours or less combined with maturity assessment can be effectively used to predict 28-day strength. Clearly, correlation between various early test results (Dhir, 1983) and 28-day strength can be achieved if sufficient data are available. Radical methods of assessing the early strength of concrete could be employed; e.g., Callander (1979) investigated the use of the Pundit (Ultra Sonic Pulse velocity) apparatus against 28-day strength. Changxiong (1981) uses rapidly determined strength and complex formulae to calculate the final strength and Kakuta (Rilem, 1991) uses sound wave propagation to predict concrete strength. When using such methods, relationships between the parameter being measured and the 28-day strength have to be developed. These types of prediction systems are normally empirical and specific to the material combinations being utilized.

The term correlation is the analysis of the differences between the predicted 28-day strengths and the actual 28-day strengths. Correlation when applied to readymixed concrete Cusum systems must not be confused with the correlation coefficient found with regression analysis. In the commercial system, the Cusum of the 'correlation' is created by comparing the actual 28-day strength with the predicted strength and then using a cumulative summation technique on the differences. Again using the data given in Table 9.4 the data in Table 9.6 would be a typical correlation calculation.

Table 9.6 Example of Cusum C to control errors from Predicted Strength and Actual Strength of test cubes

Predicted 28-day strength (MPa)	Actual 28-day strength (MPa)	Difference between Actual & Predicted Strengths	Cusum of differences Cusum C
38.0	39.0	+ 1.0	+ 1.0
43.0	44.5	+ 1.5	+ 2.5
34.5	34.0	– 0.5	+ 2.0
36.5	39.0	+ 2.5	+ 4.5
47.0	48.0	+ 1.0	+ 5.5
43.0	43.5	+ 0.5	+ 6.0
37.5	37.5	0.0	+ 6.0
40.0	41.0	+ 1.0	+ 7.0
28.0	27.5	– 0.5	+ 6.5
33.5	34.0	+ 0.5	+ 7.0

From the cumulative sum column, it is clear that the predicted strength is tending to be lower than the actual strength. As with Cusum M and Cusum R the correlation is monitored for significant changes by plotting a graph and applying a truncated V mask to the plot of the same basic design as the Cusum M graph. The action arms of this mask relate to the standard deviation of the difference in actual and predicted cube strengths and not the Cusum TSD. This is effectively the variation associated with the testing regimes and therefore much lower than the TSD. Therefore, the mask has only three lines, 'low', 'medium' and 'high' corresponding to SDs of 2, 2.5 and 3.25 MPa respectively.

When a change in correlation is detected, it is necessary to recalculate the other Cusums, e.g. M and C. This is very tedious by hand and computerization does assist. Alternatively the author has discovered that a continually updating system, using the rolling mean of ten strength differences between predicted and actual 28-day strengths, works equally as well. This is easier to computerize and prevents recalculations being necessary. Figure 9.8 shows the Cusum plots for the example calculations given above.

9.4.6 Properties of the Cusum system of quality control

The sensitivity to detecting changes in the strength parameters is a function of the shape of the mask applied. Figure 9.7 indicates the response of the traditional Cusum M (and Cusum C) masks. As these relate to normal distributions, they are identical for both upward and downward changes. However, as the distribution for standard deviation is log normal, when using a mask that has mirror image action arms the response is inconsistent between upward and downward changes, that is downward changes in SD take longer to detect. Figure 9.9 shows the response of the system.

Sear (1996) calculated a modified mask that has improved response to downward changes. This has the form for downward changes in standard deviation;

- Decision interval = 7.3 × Target Range
- Gradient of action arms = Target Range/14

Normally run lengths are calculated by computer simulation. The mean run lengths are easily calculated by resolution of the formulae for the masks. The confidence limits in general are wide and are practically impossible to calculate by hand. Computer simulation

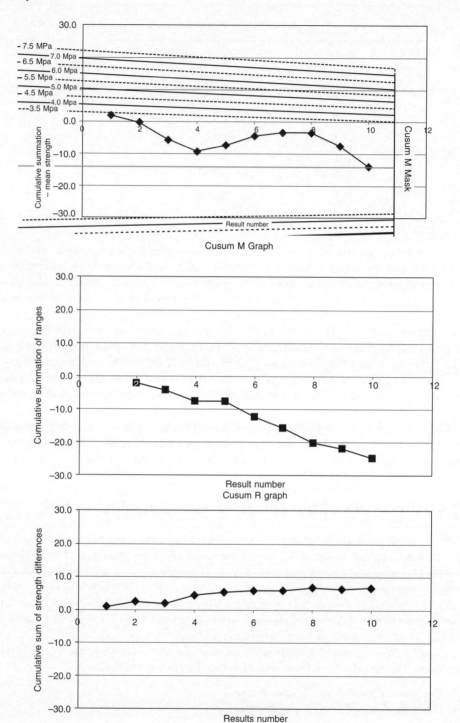

Figure 9.8 Cusum plots.

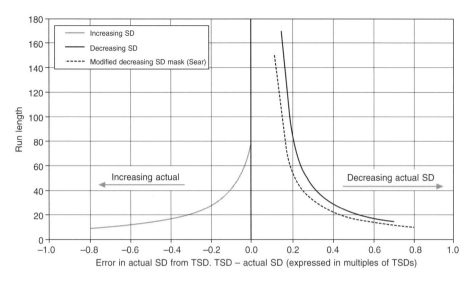

Figure 9.9 Run length for Cusum R.

becomes the most straightforward way to determine these limits. However, great care has to be taken with such forms of analysis for the random number generators used within computers are *not* truly random.

9.4.7 Cusum charts in practice

Figure 9.5 showed a large number of results plotted on a Shewhart chart. As stated, this data had a significant change in mean strength and standard deviation, which is difficult to see from the figure. However, plotting the data using the Cusum method produces charts that clearly show when these changes occurred as shown in Figure 9.10.

Visually, it is clear from these charts that the mean strength was decreasing relative to the overall mean from results number 9 to 19 and recovered from results number 20 to 40. In fact the mean strength was increased from result number 20 onwards by 6.5 MPa. Cusum M clearly shows this change in performance. For the Cusum R plot a large increase range of pairs with result numbers 36, 37 and 38 is seen. This may or may not indicate a trend – a single result (result number 36) can cause such an effect as each result is used twice in the range of pairs calculation. In fact, the dataset had the standard deviation increased by 2.5 MPa from result number 20 onwards.

While the Cusum charts do give a better visual representation of the changes in the quality of a product some care has to be taken. Using the V masks does simplify the decision making, i.e. when action is required. Changes in mean strength are in general straightforward. However, changes in standard deviation are more difficult to detect and interpret. The standard V masks tend to overestimate the standard deviation by 1.5 MPa on average.

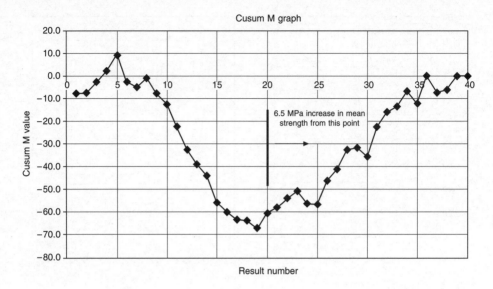

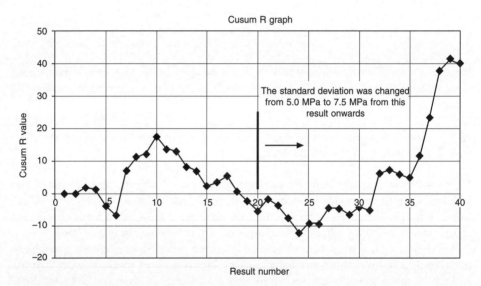

Figure 9.10 Cusum charts for the data shown in Figure 9.5.

9.4.8 The implications of taking action

For a given concrete mix, taking action can have many commercial implications. For a given change in strength or standard deviation, the mean strength of the concrete being produced should be adjusted. This is normally achieved by increasing or decreasing the cement content or water-reducing admixture dosage as appropriate. Where a change in mean strength is simply compensated for, a change in standard deviation has greater implications. As characteristic strength is governed by the margin, e.g. normally two × TSD, an increase in SD has twice the impact. Approximately 5 kg/m^3 equate to 1 MPa.

Portland cement is typically some five times more expensive than aggregate per tonne. This can have serious consequences on mix costs. Typically a 10 kg/m^3 increase (equivalent to reduction of 2 MPa on mean strength or an increase of 1 MPa for standard deviation) in Portland cement content would increase mix costs by 1.8 per cent per cubic metre. As many readymixed concrete suppliers will be operating on profit margins as low as 4 per cent this can represent a significant reduction in profits, hence quality control is highly important to the concrete producer.

9.5 Compliance or acceptance testing

The control of concrete by the various statistical methods described relies on a large number of results to ensure compliance. However, the concrete consumer is unlikely to have access to such a large number of data, especially at the start of a contract. In many standards, simplified criteria are given in order that the consumer may assess compliance using a fewer number of results. However, such simplistic criteria can lead to the consumer rejecting concrete that does comply, because of the inevitable random statistical fluctuations. Within BS EN206-I the compliance of concrete is solely the responsibility of the producer and a set of criteria for both initial and continual assessment are described. The same criteria could be used by the consumer. However, as concrete produced to BS EN 206-1 is generally third party accredited; perceived differences in compliance between the producer and consumer shouldn't arise. Nevertheless, it is interesting to look at the relationships between a simple conformity assessment system and full statistical control.

Normally, simplified conformity assessment systems often involve comparing averages from a group of results against defined values. As an example 8S5328: 1990 (BSI, 1990) used such a compliance system as shown in Figure 9.11. The compliance with the characteristic strength would be assumed if the conditions given are satisfied

The rules for C20 and above concrete were designed to operate on a discrete group of four results rather than, overlapping groups as in Table 9.7. However, the use of discrete

Table 1. Characteristic compressive strength compliance requirements			
Specified grade	Group of test results	A	B
		The mean of the group of test results exceeds the specified characteristic compressive strength by at least: N/mm^2	Any individual test result is not less than the characteristic compressive strength less: N/mm^2
C20 and above	first 2	1	3
	first 3	2	3
	any consecutive 4	3	3
C7.5 to C15	first 2	0	2
	first 3	1	2
	any consecutive 4	2	2

Figure 9.11 Table 1, BS 5328 Part 4: 1990.

Table 9.7 Differing interpretations of BS 5328 compliance rules

Results number	Discrete group of four	Overlapping groups of four				
1	1st ↑	1st ↑				
2	1st	1st	2nd ↑			
3	1st	1st	2nd	3rd ↑		
4	1st ↓	1st ↓	2nd	3rd	4th ↑	
5	2nd ↑		2nd	3rd	4th	5th ↑
6	2nd		2nd ↓	3rd	4th	5th
7	2nd			3rd ↓	4th	5th
8	2nd ↓				4th ↓	5th ↓
9	3rd ↑		Etc.			
10	3rd		Etc.			
11	3rd		Etc.			
12	3rd ↓		Etc.			

groups was usually ignored. Therefore, we shall look at the response of these rules when overlapping groups of four are used.

Operating the compliance criteria gives a significantly reduced probability of complying because low results are used four times artificially depressing the response of the criteria. The probability of concrete being rejected using such compliance criteria in comparison with the actual failure rate is known as the Operating Characteristic curve or O–C curve.

9.6 Operating-Characteristic (O-C) curves

The O-C curve is simply a graphical explanation of how discriminating the compliance criteria are. It shows the probability that a batch will be rejected or accepted as in Figure 9.12.

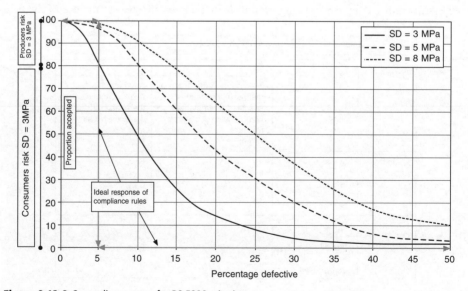

Figure 9.12 O-C compliance curve for BS 5328 criteria.

It would appear that producing concrete at a low standard deviation from Figure 9.12 is a disadvantage to the producer as there is a greater probability of concrete being rejected. However, these effects can be compensated for by applying a slightly higher margin. There is still sufficient incentive to achieve lower standard deviations. However, the unscrupulous supplier can operate at a lower margin than two × SDs with a high SD without detection! Figure 9.13 indicates the required margin for various acceptance proportions.

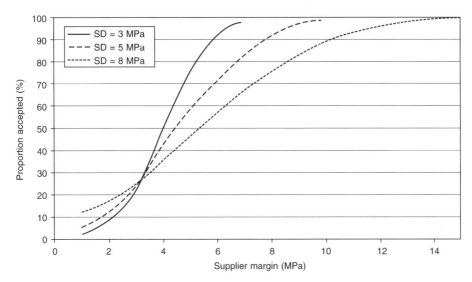

Figure 9.13 A lower standard deviation requires a greater margin to ensure acceptance.

From Figures 9.12 and 9.13 it is clear such simple rules for compliance testing lead to a situation where there is little incentive to produce consistent low standard deviation concrete. However, in order to counteract such statistical quirks one has to construct ever more complex compliance criteria. With the advent of the personal computer, it is quite feasible for the consumer to carry out detailed statistical analysis. This requires a considerable number of results before it is clear whether compliance is being achieved. From this premise, the consumer has to assess the risk that the quantity of concrete supplied is substandard and to whether this risk is acceptable.

9.7 Producer's and consumer's risk

The producer's and consumer's risk is shown on Figure 9.12 for a SD = 3 MPa at a 5 per cent failure rate. The producer has a risk of ~20 per cent of acceptable concrete being rejected, whereas the consumer has ~80 per cent chance of accepting concrete which is *not* in full compliance. As indicated before, the unscrupulous supplier with a high SD could use the compliance rules in his favour. Figure 9.12 also shows the ideal response, which is impossible to achieve in reality.

The use of overlapping groups of four leads to a situation where a low cube result will be used repeatedly, e.g. four times in any analysis. This results in knots of non-compliance in the response to the rules. However, the sampling rate will govern how much concrete is at risk. For example, if samples are taken every 10 m³ with a batch size of 2 m³, i.e. from 5 batches of concrete, the probability of accepting concrete which is substandard can be computed from the O-C curve. For this example the quantity of concrete at risk of rejection, when samples are taken randomly, is given in Table 9.8.

Table 9.8 Quantity of concrete at risk

Rate of sampling	Rate 1	Rate 2	Rate 3
	Sample from one batch selected randomly to represent an average volume of not more than:		
	10 m³	50 m³	100 m³
Batches of 2 m³ Concrete at risk			
Minimum:	24 m³	104 m³	204 m³
Maximum:	40 m³	200 m³	400 m³
Expected:	32 m³	152 m³	302 m³
Batches of 5 m³ Concrete at risk			
Minimum:	30 m³	110 m³	210 m³
Maximum:	40 m³	200 m³	400 m³
Expected:	35 m³	155 m³	305 m³

Clearly, the rate of sampling can have a significant effect on the amount of concrete at risk. The subsequent exercise to prove the concrete supplied is, or is not, satisfactory can prove very time consuming and expensive to both supplier and consumer.

9.8 Experimental design

In many ways, designing an experiment is more important than carrying out the experiment. A badly designed experiment can never be retrieved whereas a badly analysed experiment can always be reanalysed. It is important to identify the variables, e.g. strength, water–cement ratio, etc., which one wishes to study. These are called factors. As we are dealing with normal variables in most circumstances one should assess the level one wishes to detect any difference, e.g. for strength one may wish to determine differences of 2.0 MPa. Finally, we need to identify the treatment that is to be carried out, e.g. to determine the effect on strength of adding a particular plasticizer. From these three factors, we need to make a decision on the number of samples needed.

The problem of acquiring good experimental data is similar to the problems faced by a communications engineer. The receipt of a signal depends on the volume of background noise and the volume of the signal. For concrete experiments the volume of signal is analogous to the difference in strength, the noise is analogous to the testing variation and the number of results to the number of trial mixes carried out.

For a single trial mix, the number of cubes made for a given age can be calculated. The variance of mean of a number of cubes is:

- Standard Error = $SD/\sqrt{n}$ where n = number of results

- The testing variation associated with making, curing and testing concrete test cubes from plain concrete is typically 1.5 MPa. We shall use this for this example.
- The 95 per cent confidence limits for the mean value varies depending upon the number of results and the Student t value.

Table 9.9 shows that at least four cubes are required to represent a mean strength value with any reasonable level of accuracy, e.g. ~2 MPa. By using a large number of cubes this has the effect of blocking, or at least reducing, one source of variation, e.g. the testing, allowing us to make a more accurate estimate of the difference between means. We can then construct a null hypothesis test for the two means.

Table 9.9 Estimating confidence limits on the mean for a single trial mix

Number of results, i.e. the number of cubes made	Standard Error (SE) in MPa Assumes SD of testing = 1.5 MPa	95% confidence limits (two-tailed test) on the mean strength from n cubes = $\pm t_n \times$ SE Degrees of Freedom (DOF) = $n - 1$
2	$1.5/\sqrt{2} = 1.061$	$12.706 \times 1.061 = \pm 13.5$ MPa
3	$1.5/\sqrt{3} = 0.866$	$4.303 \times 0.866 = \pm 3.7$ MPa
4	$1.5/\sqrt{4} = 0.750$	$3.182 \times 0.750 = \pm 2.4$ MPa
5	$1.5/\sqrt{5} = 0.671$	$2.776 \times 0.671 = \pm 1.9$ MPa
6	$1.5/\sqrt{6} = 0.612$	$2.571 \times 0.612 = \pm 1.6$ MPa
7	$1.5/\sqrt{7} = 0.567$	$2.447 \times 0.567 = \pm 1.4$ MPa
8	$1.5/\sqrt{8} = 0.530$	$2.365 \times 0.530 = \pm 1.3$ MPa
9	$1.5/\sqrt{9} = 0.500$	$2.306 \times 0.500 = \pm 1.2$ MPa
10	$1.5/\sqrt{10} = 0.474$	$2.262 \times 0.474 = \pm 1.1$ MPa

We have estimated the standard deviation for the testing variation as shown in Table 9.9; is this a safe assumption when comparing results from trial mixes? The following example shows how sample variation can be compared using the F test.

Suppose we set up an experiment to determine whether a brand of plasticizer has a significant effect on strength when added to concrete made at a fixed workability. Will the addition of the admixture reduce or increase the variation due to testing? Two trial mixes are planned, one without the plasticizer and the other at the same slump but containing a standard dose of the admixture. The results are shown in Table 9.10 and Figure 9.14.

Table 9.10 Results from two trial mixes to compare the effect of a plasticizer

Cube result no.	Strength @ 28 days – no plasticizer	Strength @ 28 days – with plasticizer
1	34.5	40.4
2	33.1	42.7
3	35.4	42.5
4	36.9	40.3
5	36.8	38.9
Mean Overall Mean y_m = 38.162	35.3 y_{m1}	41.0 y_{m2}
Standard Deviation S_p	1.61 S_1	1.61 S_2
Square sum of differences SS^2	10.32 SS_1^2	10.31 SS_2^2

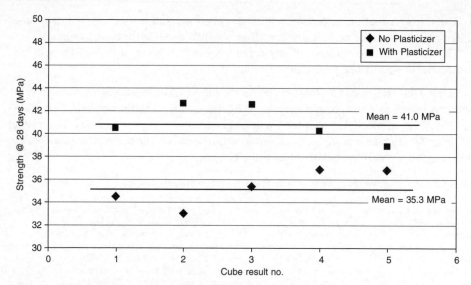

Figure 9.14 Two trial mixes – is there a real difference in strength?

We can see a 5.7 MPa strength difference between the two trial mixes. The confidence limits for the strength difference are as follows:

Strength difference with confidence limits = $(y_{m1} - y_{m2}) \pm t\alpha_{/2} \sqrt{S_w^2 (1/n_1 + 1/n_2)}$ (9.2)

where within-sample variation $S_w^2 = \dfrac{(n_1 - 1)S_1^2 + (n_2 - 1)S_2^2}{n_1 + n_2 - 2}$

n_1 and n_2 = number of results in datasets 1 and 2 respectively
$t\alpha_{/2}$ is based on $(n_1 + n_2 - 2)$ degrees of freedom (DOF) = 8
$\alpha = 0.025$ in each tail or 95% confidence

∴ $S_w^2 = (4 \times 1.61^2 + 4 \times 1.61^2)/8 = 2.592$ $S_w = 1.605$

$t\alpha_{/2} = 2.306$

∴ Mean strength difference = $5.7 \pm 2.306 \times 1.61 = 5.7 \pm 3.7$ MPa

It is clear from the calculations that the strength difference is significant. It shows the plasticizer is working producing between 2.0 and 9.4 MPa improvement in concrete strength.

Alternatively, we can check the result by comparing the within-sample and between-sample variations and carrying out an F test. The between-sample variation is the weighted sum of the squares of deviations of the individual sample means about the mean for all 10 observations, y_m, divided by the number of samples, i.e. two trial mixes. The calculation is:

Between-sample variation $S_b^2 = \dfrac{n_1(y_{m1} - y_m)^2 + n_2(y_{m2} - y_m)^2}{2 - 1}$ (9.3)

$= \dfrac{5(35.1 - 38.162)^2 + 5(41.0 - 38.162)^2}{1} = 87.15$

∴ $S_b = 9.33$

From these results, we can test the null hypothesis using the F test by comparing the between-sample variation and the within-sample variation:

$$F = S_b^2 / S_w^2 \text{ where the DOF are } v_1 = 1 \text{ and } v_2 = n_1 + n_2 - 2 = 8 \tag{9.4}$$

$F = 87.15/2.592 = 33.62$ (the F value from the 95% confidence limit table = 5.32)

∴ As the calculated F is significantly greater than the tabular value we reject the null hypothesis, i.e. the means *are* significantly different.

The above example is based on simple variables, e.g. the addition of a plasticizer. Carrying out two trial mixes with five test cubes being made from each is statistically valid. However, should more variables be included, e.g. by varying the dosage of plasticizer or by using more than a single technician or even laboratory, calculating the number of samples required and the analysis of the resulting data becomes ever more complex.

References

Barber, P. and Sym, R. (1983) An assessment of a target value method of quality control. ERMCO 83 Working Sessions W13(1), May.
Brown, R.V. (1984) Concrete Society Digest No. 6 *Monitoring Concrete by the Cusum System*, Cement and Concrete Association, Slough.
BSI (1983) BS 1881: Part 112: Methods of accelerated curing of concrete test cubes. British Standards Institution, London.
BSI (1990) BS 5328: Part 4: Concrete: Part 4. Specification for the procedures to be used in sampling, testing and assessing compliance of concrete. British Standards Institution, London.
BSI (2003) BS 5703: Parts 1 to 4. Guide to data analysis and quality control using Cusum techniques.
Callander, I.A. (1979) A Study & Investigation of UPV for the quality control of concrete. CGLI Advanced Concrete Technology project, April.
Changxiong, C. (1981) Application of rapidly determined strength and cement water ratio to quality control and assessment of concrete. *Quality of Structures, RILEM proceedings*, June, Gwent.
Day, K.W. (1995) *Concrete Mix Design, Quality Control and Specification*, E & FN Spon (Chapman & Hall), London, 110 and 140 to 147.
Dewar, R. and Anderson, R. (1988) *Manual of Readymixed Concrete*, Blackie & Son. Ltd, Glasgow.
Dhir, R.K. (1983) Concrete quality assessment rapidly and confidently by accelerated strength testing. ERMCO 83 Conference.
Dhir, R.K. and Gilhespie, R.Y. (1983) Concrete quality assessment rapidly and confidently by accelerated strength testing. ERMCO 83 Working sessions W14A (A), May.
ERMCO (1983) Variability of cements and adjustments in concrete mixed by accelerated testing. ERMCO 83 Working sessions W14A (3), May.
Francis, K. (1997) *Using the cusum system to control the quality of concrete*. Readymixed Concrete Association of Singapore.
Grant, N.T. (1977) *A Cusum controlled accelerated curing system for concrete strength forecasting*. RMC Technical services report no. 79 & ERMCO.
Jones, R. (1992) Decision rules for cusum charts. *Quality Forum*, **18**, No. 3, Sept.
Lapinas, R.A. (1975) Accelerated concrete strength testing by the modified boiling method: A concrete producers view. ACI Detroit.
Mendenhall, W. and Sincich, T. (1988) *Statistics for Engineering and the Sciences*, 3rd edn, Maxwell Macmillan International Editions, London.
Quality Scheme for ReadyMixed Concrete (QSRMC) (1991) Technical Regulations. Appendix on the Cusum control system.

Readymixed Concrete Ltd (1969) Standard technical procedure – Monitoring test variability, July.

RILEM (1991) Evaluation of very early age concrete using wave propagation method. *Quality of structures, RILEM proceedings*, June, Gwent.

Sear, L.K.A. (1996) *The development of a decision support system for the quality control of readymixed concrete*. PhD thesis.

Sweeting, I.O. (1977) Accelerated testing of high strength concrete and concrete containing plasticisers using the Grant method. CGLI Advanced Concrete Technology Project Report, May.

Woodward, R.H. and Goldsmith, P.L. (1963) *Cumulative Sum Techniques – Monograph No. 3*. Published for Imperial Chemical Industries by Oliver & Boyd.

Further reading

British Standards Institution (1990) BS 5328: 1990, Concrete: Part 4. Specification for the procedures to be used in sampling, testing and assessing compliance of concrete, BSI, London.

Mendenhall, W. and Sincich, T. (1988) *Statistics for Engineering and the Sciences*, 3rd edn, Maxwell Macmillan International Editions, London.

10

Statistical analysis techniques in ACT

Stephen Hibberd

10.1 Introduction

Concrete is notable as a material whose properties can vary widely depending on the choice and proportions of aggregates, cement, water, additives etc., together with the production technique. An associated feature is that even with a desired (target) mix the inherent variability in the materials and the production process will inevitably result in a final product that differs from the target requirement. Of course, provided the concrete remains within specified tolerances on key attributes then the product is acceptable. Effective management of concrete therefore must include a quantitative knowledge of the key attributes, monitoring techniques, decision methods, their limitations and an ability to interpret the measured values. Statistical techniques are consequently used extensively to understand and compare variations between concrete batches, to modify and control the production of concrete and to form the basis of Quality Control and Quality Assurance.

Within a short chapter it is impossible and not appropriate to provide an in-depth coverage of statistical theory and the underpinning ideas from probability theory. An emphasis is on statistical techniques that are exploited and applied explicitly to current practical circumstances in ACT such as trends and errors, estimation of parameters, checking test results, mix design, compliance and quality control. Initially each section will concentrate on providing fundamental understanding and competence of the background techniques that will be required and use will be made of relevant statistical tables and formulae.

10.2 Overview and objectives

The theory is divided into coherent sections that will provide a theoretical background to the applications of procedures covered in other chapters as follows:

- **Sample data and probability measures** This section aims to consolidate knowledge on the calculation of sample statistics and their relevance together with an understanding of probability typified by a normal distribution. Objectives are to:
  - consolidate terminology and calculation of sample statistics;
  - introduce concepts and measurements of probability;
  - evaluate probabilities using the normal distribution.
- **Sampling and estimation** Variations associated with the process of sampling are addressed and quantitative measures introduced. Two statistical distributions, the t-distribution and the F-distribution, are introduced and their use to provide estimates of key population parameters explained. Objectives are to:
  - understand the concept of sampling to provide an estimate of a key (population) values;
  - calculate of the precision of estimates for large sample sizes using the normal distribution;
  - evaluate estimates and precision for small sample sizes using the t- and F-distributions;
  - calculate confidence intervals and understand their application in constructing control charts.
- **Significance tests** The concept of decision making based on sample data is covered and applied to the comparison of mean and variances of key parameters. Objectives are to:
  - understand hypothesis testing in using sample data to test the validity of a statement;
  - understand the relationship between the significance level and critical values in tests;
  - evaluate a sample mean with a target (population) value;
  - compare target means or variances from two sets of sample data.
- **Regression models** An examination of possible relationships between linked parameters and the derivation of useable functional relationships will be covered in this section. Objectives are to:
  - understand the concept of correlation as a measure of association between sets of data;
  - calculate a 'least-squares' linear regression line;
  - understand and calculate correlation coefficients and residuals.
- **Statistical formulae and tables**
  A collection of some relevant formulae and tables used in ACT are provided.

10.3 Sample data and probability measures

10.3.1 Random variation

Within the physical world many natural quantities are subject to an amount of random variation in their formation and consequently provide variation in any materials for which they are a constituent part. In concrete technology random variations also occur due to changes in processing, for example due to minor chemical inconsistencies, mixing time variation and small water quantity changes. Thus, even with the most careful of measuring constitutive quantities, natural variations will occur in the properties of the resultant material. Random variations will also affect measurements of all quantities X, say, as these are subject to errors; if careful measurements are repeated or different instruments used then the values of X will lie close to some precise value but some discrepancy must be expected. Information and subsequent analysis on such variations can be obtained from a study of data collected from laboratory or on-site tests. Random variation is then often plotted in histogram form to identify principal characteristics such as central tendency, variability and shape. For quantitative analysis then an associated probability distribution for characterizing the variation in X is sought.

Random variations must be clearly distinguished from systematic variation that may arise from some planned change to the process or some time-varying process. For example, in the case of comparing the increased 7-day strength of concrete samples, as a result of increasing additive, then a plot of strength against quantity of additive will follow an anticipated (systematic) curve, while variation about this plot would be random error.

10.3.2 Sample data

Statistics involves dealing with information from collected data. Clearly it is important that a sufficient quantity and the correct type of information is gathered to make predictions reliable. These more advanced topics are covered later in this chapter; initially we concentrate on the key ideas associated with sampling and the representation and interpretation of data.

The most common type of 'experiment' involves taking a sample from a population, i.e. a selection of items from a whole. Ideally, the whole population would be studied, but this may be impractical for two main reasons:

- Expense – the population may be too large or testing each item may be expensive.
- Destructiveness – testing may require dismantling or running to destruction.

Generally, some form of estimation, decision or prediction is made affecting the whole population by the analysis of data from just a sample. Care must be therefore be exercised to distinguish between a

- population statistic – some value associated with the whole population (i.e. usually the quantity we need to estimate);
- sample statistic – some value obtained from a sample (i.e. a value obtained from only a part of the population).

10.3.3 Representation of data

Statistical data, obtained from surveys, experiments or any series of measurements are often so numerous that they are virtually useless unless condensed or reduced to a more suitable form. A necessary first step in any engineering situation is an investigation of available data to assess the nature and degree of variability of the physical values. Sometimes it may be satisfactory to present data 'as they are', but usually it is preferable or necessary to group the data and present the results in tabular or graphical form. For subsequent calculations or direct comparisons then some quantitative measures of data may be required. An unorganized list of data values is not easily assimilated. However, there are numerous methods of organization, presentation and reduction that can help with data interpretation and evaluation.

Histograms

Given a set of recorded data values it is useful to group the frequency of occurrence within suitable intervals. Useful histograms can be based on

- absolute frequencies – the number of data values within each interval;
- relative frequencies – the proportion of total number of data values within each interval;
- cumulative absolute frequency – the running total of absolute frequencies;
- cumulative relative frequency – the running total of relative frequencies.

Typical histograms are shown in Figures 10.1(a) and 10.1(b) corresponding to representations of data obtained from laboratory test for determining tensile strength of a concrete mix as given in example *Case 1*. The distribution of data values are made relative to intervals of width 20 and centred at mid-point values (class marks) as displayed. Figure 10.1(a), shows a typically characteristic random variation of many quantities in ACT; sample values are found to vary around some 'central location' and with some element of 'spread'.

Case 1 A sample of test results of the splitting tensile strength on 50 concrete cylinders

320	380	340	410	380	340	360	350	320	370
350	340	350	360	370	350	380	370	300	420
370	390	390	440	330	390	330	360	400	370
320	350	360	340	340	350	350	390	380	340
400	360	350	390	400	350	360	340	370	420

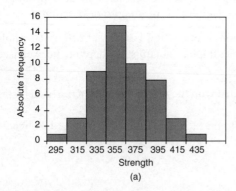
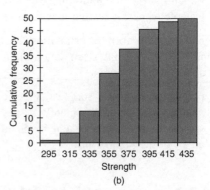

Figure 10.1 (a) Frequency histogram, (b) Cumulative frequency histogram.

Scatter diagram (Scattergram)

In measuring systematic variations of quantities X and Y, say, the recorded data values will be in the form of data pairs (x_i, y_i). An example is given in *Case 2* that provides a set of data pairs related to measured deflections from a loaded concrete beam. When plotted on a Cartesian plot (scattergram), such as Figure 10.2, any functional relationship between the two variates may become more evident. The plot in Figure 10.2 perhaps suggests a linear relationship exists between deflection and load, with variations from an exact straight-line relationship reflecting the error associated with the data.

Case 2 The measured values of beam deflections y_i against applied loads x_i

i	1	2	3	4	5	6	7	8	9	10
x_i	100	110	120	130	140	150	160	170	180	190
y	45	52	54	54	62	68	75	75	92	88

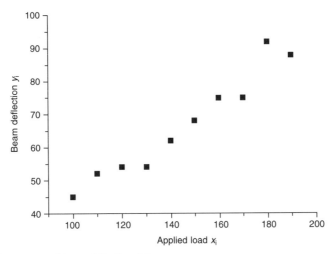

Figure 10.2 Scattergram of beam deflection data.

10.3.4 Quantitative measures

For calculation, decision making and comparison purposes it is useful to obtain standard analytic measures of the data characteristics. The two principal measures are first, that of location, given by a single representative value locating the 'centre' of the data, and second, a measure of spread or variation of data values, usually relative to the 'centre'. Given n data values, labelled x_i, say, $i = 1, n$, a measure of data values location based on the arithmetic mean of all n data values x_i is the sample mean $\bar{x}$ given by

$$\bar{x} = \frac{1}{n} \sum_{i=1}^{n} x_i$$

A measure of the spread in data values from their mean value is given by the sample variance s^2 given by

$$s^2 = \frac{1}{n-1} \sum_{i=1}^{n} (x_i - \bar{x})^2$$

and the derived value $s = \sqrt{s^2}$ termed the sample standard deviation; the latter has the same dimensions as the data values and is more frequently used in operational formulae.

The above measures are widely used but simpler measures may sometimes be more appropriate in some cases. An alternative measure for the 'centre' of the data is the median, determined from data arranged in ascending order as the middle value (odd number of data points) or the average of the middle two data values (even number of data values). The use of ordered values can be extended to give quartiles – i.e. data values divided in quarters or even finer divisions of tenths termed deciles. A simple measure of spread is given by the difference between the largest and smallest data values and called the sample range, some care is needed with this measure as it can be severely affected by a rogue data value. *Case 1* gives example data values used to obtain values of the sample mean = 363.8, sample variance = 832.2 and sample standard deviation = 28.8. By comparison the sample median = 360 and the sample range =140.

10.3.5 Population values

The above measures for mean and variance can also be applied to population values but with a significant minor alteration, for small samples in particular. For populations of a random variate X with finite number of discrete data values N, the (population) mean μ_x and the (population) variance σ_x^2 are given by

$$\mu_x = \frac{1}{N} \sum_{i=1}^{N} x_i \quad \text{and} \quad \sigma_x^2 = \frac{1}{N} \sum_{i=1}^{N} (x_i - \mu)^2$$

If no ambiguity exists with the variate referenced then these may be simply written as μ and σ^2. The difference in the dividing factor in the formula between the sample variance s^2 and population variance σ^2 arises because the mean μ is exact whereas in a sample of data values then the sample mean $\bar{x}$ provides only an estimate for the actual mean μ. The population standard deviation is defined by $\sigma = \sqrt{\sigma^2}$ as might be expected.

10.3.6 Probability

A probability is a proportional chance of a particular occurrence. Perhaps the most quoted mathematical source of probability is the throwing of a single common dice. The outcome of one throw is one of six numbers 1 to 6, each with an equal chance of occurrence, i.e. the probability of obtaining a particular score is one in six – expressed as 1/6. Mathematically an event A, say, will have

$$\text{Probability of } A = P(A) = \frac{\text{Number of outcomes which result in } A}{\text{Total number of outcomes}}$$

If events are not equally likely then an appropriate modification with appropriate weightings needs to be used; the associated link between probability values and the possible outcomes is called a probability distribution. It is evident that a probability is a numerical measure that lies between 0 and 1; if the outcome is impossible then it will have a probability of 0 while a certainty will take a value of 1. In applying probability concepts to concrete technology it is usually not so straightforward as the example above to enumerate the probability values, but these often exist as a result of past experience and expertise or on the basis of laboratory testing. Probability thus provides a theory in which the uncertainty is known (or assumed known) to follow a specific probability distribution. In dealing with probability theory we will be looking at assigning proportional chances to random events through studying their associated probability distributions.

10.3.7 Probability functions

The quantities of most interest in ACT, for example compressive strength, tend to take continuous values x, of the variate X say, and lie within a range $0 < x < \infty$. However, the chance of realizing values will be associated with some underlying probability measure. In Figure 10.1(a), a distribution of tensile strengths as obtained from testing is displayed within a histogram and indicates that values falling within different intervals had differing frequencies, i.e. different probabilities of occurrence. The corresponding histogram Figure 10.1(b) identifies how the occurrences are accumulated for increasing values of the variate X.

To generate quantities in *Case 1* consistent with probability measures then frequencies need to be scaled relative to the total sample size, i.e. to graph relative frequencies rather than absolute frequencies as shown in Figures 10.3(a) and 10.3(b). In this case the area under the histogram in Figure 10.3(a) will sum to unity and the corresponding cumulative sum in Figure 10.3(b) will approach the total value of unity.

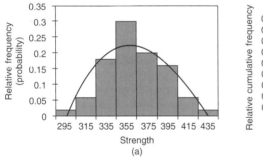
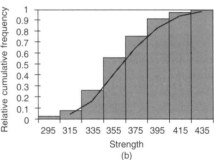

Figure 10.3 (a) Probability density, (b) probability distribution.

Although *Case 1* corresponds to a finite sample, it is straightforward to recognize that with increasing sample size, then further but smaller class intervals can be readily defined, and the histogram columns will mark out a more continuously varying area beneath characterizing continuous curves, such as those shown illustrated in Figure 10.3. This process correspondingly defines two functions in dealing with continuous variates: first, the probability density function (pdf) $f(x)$ showing the variation of probabilities for values x of the variate X and second, the cumulative distribution function (cdf) $F(x)$ that provides a summation (integration) of the probability measures for increasing values of x. These curves characterize the underlying random processes and play a crucial role in evaluating probabilities and quantifying the statistical analysis. Fortunately, a number of common distributions exist that match the characteristics of processes found in ACT situations, but even then, values are not readily obtainable from simple analytic functions but need to be evaluated from tables of values or reference to a computer-based package such as Excel or a specialized statistical package.

The cdf is given in terms of the integral of the pdf and, conversely, the derivative of the cdf is the pdf i.e.

$$\frac{dF(x)}{dx} = f(x) \quad \text{and} \quad F(x) = \int_{\text{least } x}^{x} f(x)\,dx$$

Following from our earlier work, the total probability is 1, which corresponds to the total area under the pdf curve that provides an essential constraint on the possible forms for $f(x)$. The cdf is adding probability values in the direction of increasing possible values x, so the curve will increase continually to a final value of unity. The probability value for a random variate X to lie between two values a and b, say, is the area under $f(x)$ between $x = a$ and $x = b$; evaluation is obtainable from integration of the pdf using the cdf as follows:

$$P(a < X < b) = \int_{x=a}^{x=b} f(x)\,dx = \int_{x=0}^{x=b} f(x)\,dx - \int_{x=0}^{x=a} f(x)\,dx = F(b) - F(a)$$

These ideas are crucial to evaluate relevant probability values from values of the cdf as given in tables.

10.3.8 Expected values

The pdf gives detailed information on the range of values a variate might take and their appropriate chance of occurrence. It remains useful to calculate some key quantities associated with these distributions such as the most likely (mean) value or the expected measure of spread. In this instance, all possible values are available and we are dealing with population quantities. Formally, these are called expected values and can be formulated for any function $\phi(X)$, to give a weighted average value and defined by

$$E\{\phi(X)\} = \int_{\text{all } x} \phi(x) f(x)\,dx$$

Expected values of the variate X or any powers of X are termed moments; the most important are:

(i) $\phi(X) = X$, to give the population mean μ;
(ii) $\phi(X) = (X - \mu)^2$, to give the population variance σ^2.

For any pdf function then these can be evaluated and are used within the subsequent statistical analysis in comparing the predicted population values with the observed sample values.

10.3.9 Normal distribution

The normal distribution (or Gaussian distribution) is the single most important and widely known distribution in engineering and plays a central role in the theory associated with concrete technology. The pdf is given by

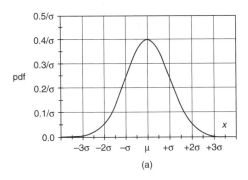
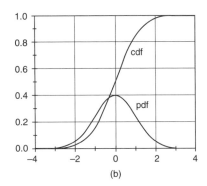

Figure 10.4 (a) pdf of $N(\mu, \sigma^2)$, (b) pdf and cdf for $N(0, 1)$.

$$f(x) = \frac{1}{\sigma\sqrt{2\pi}} \exp\left\{-\frac{(x-\mu)^2}{2\sigma^2}\right\} \quad \text{with} \quad -\infty < x < \infty$$

This involves two characterizing parameters corresponding to the mean μ and standard deviation σ of the distribution and accordingly the distribution is denoted by $N(\mu, \sigma^2)$ for convenience.

The pdf has a bell-shaped curve as shown in Figure 10.4(a) which is symmetric about the mean μ, a maximum value of $1/\sigma\sqrt{2\pi}$ and with a shape that rapidly decays to zero for values away from the mean. While $f(x)$ has non-zero values for all positive and negative values of the variate x, it is negligible for most practical purposes when x is more than a distance 3σ from μ. The corresponding cdf is shown in Figure 10.4(b) and although it does not have a simple functional formulation it can be evaluated numerically. Figure 10.4(b) shows both the pdf and cdf of $N(0, 1)$, i.e a distribution with mean zero and variance unity.

The distribution for $N(0, 1)$ is particularly important as it provides a base calculation for any normal distribution and is consequently well tabulated. Statistical Table 10.1 gives a table of values shown in Figure 10.4(b).

The normal distribution is often used for its relative simplicity combined with a proven ability to provide accurate quantitative information when used in appropriate circumstances. It also has the useful 'addition' property that if $X_1 \sim N(\mu_1, \sigma_1^2)$ and $X_2 \sim N(\mu_2, \sigma_2^2)$ then the random variables

$$X_1 + X_2 \sim N(\mu_1 + \mu_2, \sigma_1^2 + \sigma_2^2) \quad \text{and} \quad X_1 - X_2 \sim N(\mu_1 - \mu_2, \sigma_1^2 + \sigma_2^2)$$

i.e. that new quantities constructed from addition of normal variables will be normal and the resulting parameters can be readily calculated.

10.3.10 Calculation of probability values (from standard tables)

Probabilities associated with the standardized normal variate $N(0,1)$ can be obtained directly from a table of values for the cdf $F(z)$, and some algebraic manipulation. The basic approaches are displayed in the following examples illustrated in Figure 10.5, where the required probability measure corresponds to evaluating the area of the marked portions of the pdf; values are determined from Statistical Table 10.1. As is typical with tables for symmetric distributions, the values corresponding to only positive values of the variate X are explicitly displayed. However, corresponding negative values are readily obtained from the symmetry of the pdf as:

$$f(-z) = f(z) \quad \text{and} \quad F(-z) = 1 - F(z)$$

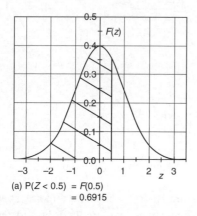
(a) $P(Z < 0.5) = F(0.5)$
$= 0.6915$

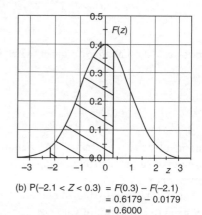
(b) $P(-2.1 < Z < 0.3) = F(0.3) - F(-2.1)$
$= 0.6179 - 0.0179$
$= 0.6000$

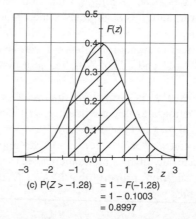
(c) $P(Z > -1.28) = 1 - F(-1.28)$
$= 1 - 0.1003$
$= 0.8997$

Figure 10.5 Evaluations of probability values from the standardized normal distribution $N(0, 1)$.

10.3.11 Standardized normal variate

The normal distribution has a number of special properties, one of which is that it has a simple scaling rule. This allows all calculations for probabilities from any normal distribution to be calculated from the single set of numerical values of $N(0, 1)$, called the standardized normal distribution, graphed as Figure 10.4(b) and values given in Statistical Table 10.1.

For any normal variate X, then an associated standardized variate Z is defined by

$$Z = \frac{X - \mu}{\sigma}$$

and can be shown to have the property that $Z \sim N(0, 1)$. Thus in practical calculation, probability values for a distribution $N(\mu, \sigma^2)$ can be rewritten and then evaluated in terms of $N(0, 1)$. As an example consider $P(X < a)$, where $X \sim N(\mu, \sigma^2)$. Then by simple algebra

$$P(X < a) = P\left(\frac{X - \mu}{\sigma} < \frac{a - \mu}{\sigma}\right)$$

i.e,

$$P(X < a) = P\left(Z < \frac{a - \mu}{\sigma}\right)$$

where

$$Z = \frac{X - \mu}{\sigma}$$

is the standardized variate and the value of this probability is readily obtained from the distribution $N(0, 1)$. In a similar way, any calculation of probabilities from a general normal distribution involving a random variate X can be recast into an equivalent calculation in terms of a standardized variate Z and calculation found in terms of a single set of normal values – the standardized normal values. Such values are either tabulated or held in any computer statistics package.

10.3.12 Example

A specification for the cement content of pavers is specified as 16.9 per cent from contractors. The mean and standard deviation of the cement contents of 50 pavers were tested as 17.2 per cent and 1.8 per cent, respectively. Contractors would be concerned if many pavers had cement contents below 15 per cent. Assuming the cement content follows a normal distribution, estimate the number of pavers that would be below standard.

Let X = cement content of pavers, then $X \sim N(17.2, 1.8^2)$. The probability of a single paver with cement content less than 15 per cent is calculated as

$$P(X < 15) = P\left(\frac{X - 17.2}{1.8} < \frac{15 - 17.2}{1.8}\right) = P(Z < -1.22)$$

Calculation of the probability is reduced to finding an associated probability of the standardized variate Z. Using Statistical Table 10.1, the probability is given by $P(Z < -1.22) = F(-1.22) = 1 - 0.8888 = 0.1112$. Thus the approximate total number of pavers = $50 \times 0.1112 = 5.56$.

The calculated number is approximately 6 pavers but there are several sources of error. Variations will exist in the measurements particularly as cement content analyses are not exact measurements of the actual contents of selected pavers. In statistical terms also, the sample mean and standard deviation used in the calculation are sample values and not population values. Assessment of accuracy of such estimates is covered in later sections.

10.3.13 Critical values

Many probability distributions identify that the variate can take a wide range of possible values but most of the probability is assigned within a relatively small range, e.g. for a standardized normal distribution approximately 95 per cent of the probability lies within a range $-2 < z < 2$. Correspondingly, untypical values of the variate are identified from the wider extremes of the pdf often called the 'tails' of the distribution. The extent of these regions depends on the specified probability to be designated in the tails. For computational purposes it is useful to identify these critical regions and enumerate these limits, called critical values, for given probabilities, called significance levels, at the extremes of the pdf. Illustrations of possible critical regions and associated critical values are shown for $N(0, 1)$ in Figure 10.6 for a significance level of $\alpha = 0.05$.

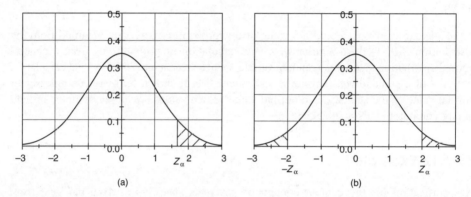

Figure 10.6 (a) Right-hand tail ($\alpha = 0.05$), (b) right- and left-hand tails ($\alpha = 0.025$ in each tail).

(a) *One tail* – Figure 10.6(a) shows the right-hand tail region for $N(0, 1)$ which contains a probability (significance level) of 0.05 (5 per cent). The critical value z_α is readily determined from Statistical Table 10.1 as $z_\alpha = 1.645$. For this symmetric distribution the critical values for an analogous left-hand tail are readily obtained as the negative of right-hand critical values.

(b) *Two tails* – Figure 10.6(b) shows the critical regions corresponding to a total probability (significance level) of 0.05 (5 per cent) divided equally into two tail regions, an amount $\alpha = 0.025$ is in *each* tail; the corresponding critical values are $z_\alpha = 1.960$ and $-z_\alpha = -1.960$.

Generally for two-tailed critical regions the probability is divided equally into both tails and significance level taken as 2α. Critical values z_α can be readily found from the cdf values of the distribution but for convenience it is useful to compile a separate table of critical values such as shown in Table 10.1 for the standardized normal distribution corresponding to typical significance levels $\alpha = 0.05$ (5 per cent), 0.025 (2.5 per cent) or 0.01 (1 per cent) in the right-hand (upper) tail.

Table 10.1 Critical value z_α for $N(0,1)$ with significance levels α (right-hand tail)

α	0.25	0.1	0.05	0.025	0.01	0.005	0.001
z_α	0.675	1.282	1.645	1.960	2.326	2.576	3.090

10.4 Sampling and estimation

10.4.1 Sample statistics

Statistical Inference is concerned with using probability concepts to quantitatively deal with uncertainty in obtaining representative values and making decisions. The basis is to obtain samples (from a population) to analyse and infer properties of the whole population. For example, to obtain the 'true' (i.e. population) average potential strength of concrete in a structure, one would need to put all the concrete for this structure into cubes and test them! Clearly this is not desirable or feasible, so an alternative is to take a number of random samples and obtain an estimate of the potential strength from the sample. For a given mix, the sample mean from testing the compressive strength of five concrete cubes might be calculated and taken as the strength of the whole mix. However, for more detailed analysis we might wish to know quantitatively the possible error with such a small sample size and to what level this might be improved by taking a larger sample. In this section details are provided of the underlying theory, probability distributions and how these are applied to use sample values to infer a value or parameter associated with the population. This may involve giving a range of values, called a confidence interval, consistent with a specified probability.

The underlying concept is that typically a collection of sample values X_i will be used to form a statistic Z, formed by some appropriate combination of sample values to provide an estimate of a population value. Each of the quantities X_i will take random values determined by its own distribution, but in estimation it is distribution of Z, the sample statistic, which is required. The most important statistics for use in ACT will be the sample mean $\overline{X}$ as an estimator for the population mean μ and the sample variance s^2 as an estimator for the population variance σ^2.

10.4.2 Large-sample statistics (normal distribution)

The most widely assumed distribution for a sample mean $\overline{X}$ is that it is normal or can be well approximated by a normal distribution. This is because many natural phenomena tend to vary symmetrically around some mean value and with variations that fall off rapidly from some mean value. Furthermore, use of the normal distribution for the sample mean when large samples are involved is justified by the following mathematical result (*Central Limit Theorem – CLT*). The result is exact for the case where sampling occurs for variables X_i that can each be assumed to vary as normal a normal distribution.

If n random samples are taken from a population with mean μ and standard deviation σ then the sampling distribution of $\overline{X}$ the sample mean will be approximately normal with mean μ and standard deviation $\sigma/\sqrt{n}$, the approximation improving as n becomes larger,

i.e. $\overline{X} \sim N\left(\mu, \dfrac{\sigma^2}{n}\right)$ or, in terms of the standardized variate,

$$Z = \dfrac{\overline{X} - \mu}{\sigma/\sqrt{n}} \sim N(0, 1)$$

This indicates explicitly that 'on average' the sample value $\bar{X}$ will predict the population value μ and that the accuracy will be characterized by an associated standard error, more explicitly written as $\sigma_{\bar{X}} = \sigma_X/\sqrt{n}$. Thus, the standard deviation of the sampling distribution $\sigma_{\bar{X}}$ is affected by both the standard deviation of each sample value σ_X but also with the number of sample values n. The effect of sample size can be assimilated graphically as shown in Figure 10.7 which shows the sampling distribution Z from a distribution with μ = 5 and σ = 1 for increasing sample sizes n = 5, 25 and 100. In each case, the mean value of Z is centred around μ = 5 but the probability associated with any individual test shows that the possibility of recording a sample value as inaccurate as 0.5 or greater from the population value is feasible for n = 5, unlikely for n = 25 but negligible for n = 100. Evaluating these probabilities using the normal tables gives respective probability values 0.2628, 0.0124, 0.0000. It also illustrates that the standard error $\sigma/\sqrt{n}$ decreases only relatively slowly, as $1/\sqrt{n}$, with increasing sample size n.

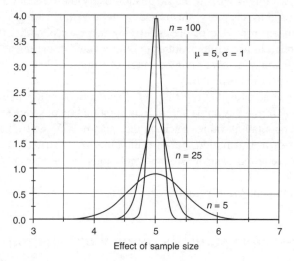

Figure 10.7 pdf of sampling distributions with n = 5, 25, 100.

The CLT result is useful provided a large enough sample is taken, often $n > 30$ is usually good enough, but in practice the population standard deviation σ is also unknown and needs to be approximated by the sample standard deviation s; this result remains a good approximation but for sufficient accuracy we may need an increased sample size to compensate of $n > 80$.

10.4.3 Small-sample statistics (*t*-distribution)

The statistical analysis provided in the previous section gives underpinning theory that can be applied in ACT but for practical purposes it is not feasible to always have large sample sizes. When a sample size n is not large then the distribution for the sample mean $\bar{X}$ is no longer accurately approximately by a normal distribution and a more appropriate distribution is the *t*-distribution. For much of the practical testing required in ACT then the appropriate statistical distributions is the *t*-distribution. The form of the probability distributions (pdf) of $t_{[2]}$ and $t_{[5]}$ as shown in Figure 10.8 compared to the pdf of the standardized normal distribution $N(0, 1)$.

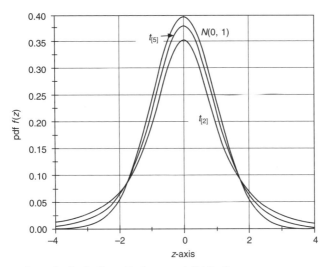

Figure 10.8 Comparison of *t*-distribution with the normal distribution.

The sampling variate for the population mean and corresponding sample distribution is defined by

$$t = \frac{\overline{X} - \mu}{s/\sqrt{n}} \sim t_{[n-1]}$$

This variate is similar to the expression from using the CLT except that it incorporates directly the sample standard deviation s to replace the usually unknown population value σ. The characterizing sample distribution is one of a family of *t*-distributions selected by a parameter $v = n - 1$, called the number of degrees of freedom. The formula for the curve of the distribution is complicated but values are tabulated in the same way as the normal distribution. Important characteristic properties of the distribution are:

- the *t*-distribution is symmetric (only positive values are usually tabulated);
- less peaked at the centre and higher probability in the tails than the normal distribution;
- a marginally different distribution exist for each value of v;
- as v becomes large, the *t*-distribution tends to the standardized normal distribution $N(0, 1)$, ($v = \infty$);
- values are obtained from *t*-tables although a restricted set of critical values often suffice.

A selection of critical values $t_{\alpha;v}$ for the *t*-distribution, corresponding to those for the normal distribution ($V = \infty$), is given in Table 10.2. A more complete listing of cdf values is given in Statistical Table 10.2.

Table 10.2 Upper critical values $t_{\alpha;\nu}$ of t-distribution $t_{[\nu]}$ for one-tailed significance level α

ν	1	2	3	4	5	6	7	8	9	10
$\alpha = 0.05$	6.31	2.92	2.35	2.13	2.02	1.94	1.89	1.86	1.83	1.81
$\alpha = 0.025$	12.71	4.30	3.18	2.78	2.57	2.45	2.36	2.31	2.25	2.23
$\alpha = 0.01$	31.82	6.96	4.54	3.75	3.36	3.14	3.00	2.90	2.82	2.76

ν	12	14	16	18	20	25	30	60	120	∞
$\alpha = 0.05$	1.78	1.76	1.75	1.73	1.72	1.71	1.70	1.67	1.66	1.64
$\alpha = 0.025$	2.18	2.14	2.12	2.10	2.09	2.06	2.04	2.00	1.98	1.96
$\alpha = 0.01$	2.68	2.62	2.58	2.55	2.53	2.49	2.46	2.39	2.36	2.33

10.4.4 Confidence intervals

Given sample values $X_1, X_2, \ldots X_n$ then using a suitable statistic Z an estimate for a population value can be obtained from a sample value, but importantly an indication of the accuracy of this estimate may also be required explicitly. Knowledge of the distribution of the sample statistic can be used to determine an interval within which the population value might lie with a specified probability. Such a prescribed probability is called the confidence level, c say, and the resulting interval the confidence interval. The confidence level may be expressed directly as a probability, e.g. 0.95, but is often expressed as a percentage, i.e. confidence level of 95 per cent.

As an example of the technique, consider obtaining the confidence interval for the population mean μ from sample values. In this instance the sample statistic Z is usually denoted by t and is given from earlier by

$$t = \frac{\overline{X} - \mu}{s/\sqrt{n}} \sim t_{[\nu]}, \quad \text{where } \nu = n - 1$$

Sample data will provide values for $\overline{X}$, s and ν. We first seek to determine an interval $-t_{\alpha;\nu} < t < t_{\alpha;\nu}$, say, that has an associated probability of 0.95 for the variate t. As the total probability is 1, then determination of values $t_{\alpha;\nu}$ is identical to determining the two-tailed critical values as shown in Figure 10.6(b) with a probability $(1 - c)$ distributed between each tail. With $c = 0.95$ then the corresponding critical values correspond to obtaining critical values associated with a significance level of $\alpha = 0.025$ in each tail. Hence the critical values are given by $t_\alpha = t_{\alpha;\nu}$ and determined from Table 10.2. Example values are:

$$n = 10 \ (\nu = 9); \quad t_{\alpha;9} = 2.25$$
$$n = 6 \ (\nu = 5); \quad t_{\alpha;5} = 2.57$$
$$n = 100 \ (\nu = 99); \quad t_{\alpha;\infty} = 1.98$$

With $n = 10$, the confidence interval is

$$-2.25 < \frac{|\overline{X} - \mu|}{s/\sqrt{10}} < 2.25$$

and can be rearranged as

$$\overline{X} - 2.25\frac{s}{\sqrt{10}} < \mu < \overline{X} + 2.25\frac{s}{\sqrt{10}}$$

to provide a confidence interval for the population mean μ with a confidence level of 95 per cent, once sample values for $\overline{X}$ and s are substituted.

Example

A sample of 32 concrete cubes from a certain mix were crushed and the average strength was 30 Nmm^{-2} with a standard deviation of 5 Nmm^{-2}. What is the 95 per cent confidence interval for the strength of the population mean for this type of mix?

The mix strength μ is determined from sample values $n = 32$, $\overline{X} = 30$, $s = 5$. The significance level in each tail is $\alpha = 0.025$, $v = 31$ and critical values are determined from Table 10.2 as $t_{0.025;31} = 2.04$ (nearest table values used) and hence a 95 per cent confidence interval is evaluated as

$$30 - 2.04 \times \frac{5}{\sqrt{32}} < \mu < 30 + 2.04 \times \frac{5}{\sqrt{32}}$$

to provide a 95 per cent confidence interval for mix strength as $28.20 < \mu < 31.80$.

10.4.5 Control charts

Confidence intervals can be exploited to provide the basis of control charts to monitor production values through ongoing sampling. For large sample sizes, critical values of the normal distribution can be used to determine corresponding confidence intervals for a population mean μ as

$\frac{|(\overline{X} - \mu)|}{\sigma/\sqrt{n}} < 1$ with probability 0.68, i.e. $\overline{X} - \sigma/\sqrt{n} < \mu < \overline{X} + \sigma/\sqrt{n}$ with 68 per cent confidence;

$\frac{|(\overline{X} - \mu)|}{\sigma/\sqrt{n}} < 2$ with probability 0.95, i.e. $\overline{X} - 2\sigma/\sqrt{n} < \mu < \overline{X} + 2\sigma/\sqrt{n}$ with 95 per cent confidence;

$\frac{|(\overline{X} - \mu)|}{\sigma/\sqrt{n}} < 3$ with probability 0.998, i.e. $\overline{X} - 3\sigma/\sqrt{n} < \mu < \overline{X} + 3\sigma/\sqrt{n}$ with 99.8 per cent confidence.

Figure 10.9 illustrates that confidence intervals can be used to obtain predetermined levels, within which sample mean values should lie with a given confidence. For example, values should lie within the lines marked 'action' with 95 per cent confidence. Thus, successive sample values from an ongoing process can be monitored and compared to the expected population value. Sample values will naturally vary from the population value but if recorded values lie outside set levels of confidence then increasingly strong indication is provided that the production system requires attention. Application of this concept can be further developed to provide Shewart and CUSUM charts as discussed in Chapter 9.

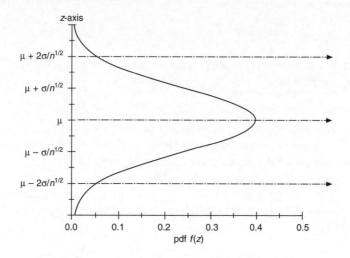

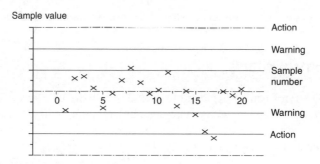

Figure 10.9 Use of confidence intervals as basis of Control Chart.

10.4.6 Comparison of means

An important technique is to be able to compare the population means between two, possibly competing, processes. This might be to ascertain, for instance, whether changing a mix formula will provide an actual increase in compressive strength. The practical difficulty is that a mean value, calculated for each process, will be obtained from sampling and so will be affected by random variation. Consequently, any difference may be accounted for entirely by natural variation and not indicating any changes within the underlying (population) values. Similarly, any measured differences in the sample values may be an over-estimate or an under-estimate of the the effect on the population values. To account for sampling then it is useful to identify a confidence level with any calculation of differences between population values.

In the case of comparing the population means μ_1 and μ_2 from separate processes X_1 and X_2 with large sample sizes n_1 and n_2 then from the CLT applied to each process the separate sample statistics for the sample means $\overline{X}_1$ and $\overline{X}_2$ are

$$\overline{X}_1 \sim N(\mu_1, \sigma_1^2/n_1) \quad \text{and} \quad \overline{X}_2 \sim N(\mu_2, \sigma_2^2/n_2)$$

involving each individual process standard deviations σ_1 and σ_2. The 'addition' property of the normal distribution can be used to obtain a sample distribution of the difference as

$$\overline{X}_1 - \overline{X}_2 \sim N(\mu_1 - \mu_2, \ \sigma_1^2/n_1 + \sigma_2^2/n_2)$$

i.e. that the variation measured from the difference in sample means is described by a normal distribution centred around the difference in population means. This result also highlights the importance of the standard error (standard deviation) of the sampling distribution for the comparison of means from two large populations, namely

$$\sigma_{\overline{X}_1 - \overline{X}_2} = \sqrt{\frac{\sigma_1^2}{n_1} + \frac{\sigma_2^2}{n_2}}$$

For calculation then the appropriate standardized normal variate is

$$Z = \frac{(\overline{X}_1 - \overline{X}_2) - (\mu_1 - \mu_2)}{\sqrt{\sigma_1^2/n_1 + \sigma_2^2/n_2}} \text{ and distributed as } Z \sim N(0, 1).$$

As previously in dealing directly with the CLT, the population standard deviation σ is generally unknown and must be approximated by the sample standard deviation s. In a similar way as earlier, this process can be adopted with little error for large sample sizes. For the practical case of comparing the population means from two general processes X_1 and X_2, irrespective of sample size, with sample sizes n_1 and n_2, then a modification to the sample statistic is possible that leads to a sample statistic involving a t-distribution. The sample statistic is

$$t = \frac{(\overline{X}_1 - \overline{X}_2) - (\mu_1 - \mu_2)}{s\sqrt{1/n_1 + 1/n_2}} \quad \text{where} \quad s^2 = \frac{(n_1 - 1)s_1^2 + (n_2 - 1)s_2^2}{n_1 + n_2 - 2}$$

and $t \sim t_{[v]}$ with $v = n_1 + n_2 - 2$.

This result also highlights the general result for the standard error of the sampling distribution for the comparison of means from two populations, namely

$$\hat{\sigma}_e = s_p \sqrt{\frac{1}{n_1} + \frac{1}{n_2}} \quad \text{where} \quad s_p^2 = \frac{(n_1 - 1)s_1^2 + (n_2 - 1)s_2^2}{n_1 + n_2 - 2}$$

Example

Data from two mixes were tested. From the first mix, 30 cubes were tested and found to have a mean strength of 38 Nmm^{-2} and a standard deviation of 3 Nmm^{-2}. The second mix provided 40 cubes with a mean strength of 36 Nmm^{-2} and a standard deviation of 2 Nmm^{-2}. Obtain a 95 per cent confidence interval for the difference in mix strengths.

Using the notation above, for the first mix (X_A) then $n_A = 30$, $\overline{X}_A = 38$, $s_A = 3$

and for the second mix (X_B) then $n_B = 40$, $\overline{X}_B = 36$, $s_B = 2$.

The parameter values for the t-distribution are $\overline{X}_A - \overline{X}_B = 2$, $v = 68$ and $s = 2.48$.

With a confidence level of 95 per cent then a corresponding two-tailed critical region is defined by

$|t| < 2.00$ following reference to Statistical Table 10.2 for the critical values for $t_{0.025;68}$

i.e. $\left|\dfrac{2 - (\mu_A - \mu_B)}{2.48\sqrt{1/30 + 1/40}}\right| < 2.00$ with a probability of 0.95.

This gives a confidence interval for the difference in mix strengths as $0.80 < (\mu_A - \mu_B) < 3.2$, i.e. that the strength of mix A is stronger than the strength of mix B by a value of between 0.8 Nmm^{-2} and 3.2 Nmm^{-2}.

10.4.7 Comparison of variances

In comparing two processes then it may be useful to compare the amount of the underlying variation induced by the random elements of each, i.e. to compare the population variances. To confirm if given samples from a population X_1 and from a population X_2 are consistent in having equal population variances $\sigma_1^2 = \sigma_2^2$ then a statistic is available as

$$F = \dfrac{s_1^2}{s_2^2} \text{ which is distributed as an } F\text{-distribution, } F_{[v_1, v_2]}$$

where $v_1 = n_1 - 1$, $v_2 = n_2 - 1$ are parameters to define each distribution.

The distribution for the F-distribution depends on two degrees of freedom v_1, v_2 and has a general shape as shown in Figure 10.10

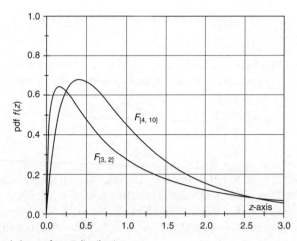

Figure 10.10 Typical shape of an F-distribution.

Important properties of the F-distribution are:

- the F-distribution is not symmetric;
- marginally different distributions exist for each pair of value v_1, v_2
- critical values are obtained from F-tables (see Statistical Table 10.3).

Within typical use, the requirement is to use a two-tailed test, i.e. to evaluate lower and upper critical values c_1 and c_2, say, corresponding to a significance level 2α, from

$$P(F < c_1) = \alpha \text{ and } P(F > c_2) = \alpha$$

In practice, it is usual to use only the upper critical value $c_2 = F_{\alpha;v_1,v_2}$, corresponding to a probability value α in the upper tail, by ensuring the sample statistic is taken as

$$F = \text{largest sample variance/smallest sample variance } (>1 \text{ automatically})$$

and readily accomplished by designating the sample data values corresponding to the larger sample variance as the numerator and the smaller sample variance data for the denominator. If required, values for the lower critical value c_1 are obtainable from Statistical Table 10.3 as $c_1 = 1/F_{1-\alpha;v_2,v_1}$; in this calculation the upper part of the distribution has probability $(1-\alpha)$. As an illustration, with $\alpha = 0.05$, $v_1 = 6$, $v_2 = 4$ then the upper critical value $c_2 = F_{0.05;6,4} = 6.16$ and the lower critical value $c_1 = 1/F_{0.95;4,6} = 1/4.53 = 0.22$. An example of the use of the F-variate is given in the next section.

10.5 Significance tests

10.5.1 Hypothesis testing

As part of monitoring manufacture or supply of components we may need to test the validity of a statement (or hypothesis) relevant to a population value by analysing a sample. Consider the illustrative example of a manufacturer of poker vibrators that claims that their product has an average life of 500 hours. Results from monitoring a sample of 36 such vibrators showed that the average life was 450 hours with a standard deviation of 150 hours. Does this disprove the manufacturer's claim?

In this case we are looking to evaluate the validity of the manufacturer's claim to within what might be regarded as a reasonable probability. Clearly, due to the variability in poker use it would not be anticipated that all pokers would last exactly 500 hours or even that a sample of 36 would have a mean value of 500 hours. If the mean value were 495 hours, say, then it might be suspected that the manufacturers claim was upheld while a value of 300 hours would raise significant concerns. It is perhaps 'reasonable' that a sample value of 450 hours is consistent with a target (i.e. population) lifespan of 500 hours, the difference being accountable to sampling variations. To provide quantitative measures to help then a statistical procedure termed 'hypothesis testing' is available that is linked to a stated probability value considered as 'unreasonable' – the significance level. In this example the random variables are

$$X_i \equiv \text{lifetime of each poker } i, \, i = 1, 2, \ldots, 36$$

and the stated (population) mean lifetime $\mu = 500$.

Given the sample of pokers, we test the hypothesis that $\mu = 500$. This is called the null hypothesis and denoted as H_0 (i.e. H_0: $\mu = 500$) and is characterized as giving a specific value in order to determine an appropriate *test statistic*. Statements such as $\mu < 500$ or $\mu > 500$ for the null hypothesis are not specific enough to formulate a subsequent analysis but do provide possible alternatives to accepting the null hypothesis. If the data is conclusive enough to reject H_0 then we must be accepting an alternative hypothesis H_1, say, which generally will affect the decision and so must always be clearly identified.

In the example there are two obvious choices for H_1

(i) $H_1 : \mu \neq 500$ – the poker manufacturer may find this favourable since
$\mu < 500$ means goods are under specification
$\mu > 500$ means specification could be upgraded.
(ii) $H_1 : \mu < 500$ – the consumer is interested in under-specification.

The sample data will be used to devise a test statistic in order to decide whether to accept or reject H_0. In this case it is an estimator for the population mean that is required as given in terms of the sample mean $\overline{X}$. The most appropriate sample statistic will be

$$t = \frac{\overline{X} - \mu}{s/\sqrt{n}}$$

with population mean μ and have a sample distribution $t \sim t_{[n-1]}$.

In Case 1 the hypothesis H_1 would be selected over the null hypothesis H_0 if $\overline{X}$ is sufficiently greater than 500 or sufficiently smaller than 500; these discriminating values are associated with two-tailed critical values. With Case 2 then H_0 would only be rejected in favour of H_1 if the sample value was sufficiently small with the discriminating critical value associated with a one-tail (left-hand) critical value.

The rationale for hypothesis testing is to assume that H_0 is true (and so the statistical analysis based on H_0 is valid) and to identify the related critical value(s) for a specified significance level α, associated with the choice of H_1. The value of the significance level is determined by circumstance or regulation but a typical value used for illustration is taken as $\alpha = 0.05$ (sometimes quoted as 5 per cent). The sample data is used to identify the validity of H_0 by checking if the sample value is consistent with the statistical analysis. This is determined by checking if the data value falls within the anticipated range (acceptance region) of probability of size $(1 - \alpha)$ as determined by the critical values. Critical values are illustrated in Figure 10.11; Case 1 involves two tail regions defined by a lower value c_1 and an upper value c_2 whilst Case 2 has an upper tail region identified by an upper critical value c_3. Values of the critical values are found directly from the appropriate table of critical values for the sample distribution. If the sample value falls outside a critical value (i.e. sample value falls within an appropriate tail region) then it is deemed not acceptable and the null hypothesis is considered untrue as a consequence.

In this example,

Case 1: $H_0 : \mu = 500$, $H_1 : \mu > 500$ or $\mu < 500$
and on taking values for μ, s and n from the null hypothesis the test statistic is

$$t = \frac{\overline{X} - 500}{150/\sqrt{36}}$$

The sample distribution is $t_{[35]}$. Taking a significance level of $\alpha = 0.05$ split between two-tails (i.e. a probability 0.025 in each tail) the critical values are readily determined from Table 10.2 as $c_1 = -2.04$ and $c_2 = 2.04$ (using symmetry of the t-distribution).

The data value for the test statistic is obtained on substituting the observed value for the sample mean as

$$\hat{t} = \frac{450 - 500}{150/\sqrt{36}} = -2.00$$

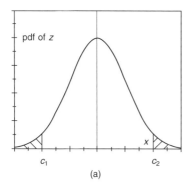
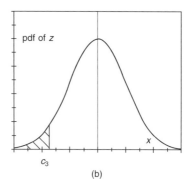

Figure 10.11 (a) Rejection regions Case 1, (b) rejection region Case 2.

This value falls within the acceptance region (marginally), i.e. within the bulk region outside of the rejection region(s). A conclusion is that, relative to the alternative hypothesis, the observed value falls within a region that is consistent with the random variation anticipated and the hypothesis $\mu = 500$ is accepted.

Case 2: $H_0: \mu = 500$, $H_1: \mu < 500$.
The test statistic and the data value is the same as Case 1, i.e

$$t = \frac{\overline{X} - 500}{150/\sqrt{36}} \quad \text{and} \quad \hat{t} = \frac{450 - 500}{150/\sqrt{36}} = -2.00$$

In this case the only rejection region for the null hypothesis is defined by a one-tailed (lower) region with significance level $\alpha = 0.05$ and determined with reference to Table 10.2 as $c_3 = -1.70$ (from symmetry upper and lower values differ only by a sign).

The observed value of the test statistic $\hat{t} = -2.00$ therefore falls within the rejection region and a conclusion is that the manufacturer's claim has not been achieved, i.e. the difference between the claimed and observed values cannot be attributed to natural variations with the stated significance level and the hypothesis $\mu = 500$ is rejected and $\mu < 500$ is accepted.

The discrepancy in conclusions between Cases 1 and 2 highlights the importance of identifying an appropriate alternative hypothesis and specification of the significance level.

Summary – general theory

The stages identified in the above example can be applied more generally to other practical situations and follow a similar approach although the choice of test statistic and its associated distribution will change accordingly. A summary of the stages is:

1. Make a null hypothesis H_0 and an alternative hypothesis H_1 (H_0 is *always* chosen to be specific to fully specify a sample statistic).
2. Assume H_0 is true and identify an appropriate test statistic t and its associated distribution.
3. Obtain a numerical value for t using the given sample data $\hat{t}$, say.
4. Specify a significance level α and determine a critical value (or values).
5. Accept H_0 or reject (i.e. accept H_1) depending on whether the sample value $\hat{t}$ falls within an acceptance or rejection region.

10.5.2 Comparison of means

Hypothesis testing can be readily used to help determine if differences on sample mean values measured between two processes are significant as shown in the following example.

Example

Concrete from two separate mixes averaged 25 Nmm^{-2} and 30 Nmm^{-2} respectively. In both cases 6 cubes were taken. Calculate if there is a significant difference between the two mixes when the standard deviation of the first was 3 Nmm^{-2} and the second was 5 Nmm^{-2}.

The variates in this case are X_1 and X_2 the strength of each mix. Associated data values for each mix are sample sizes $n_1 = 6$ and $n_2 = 6$, sample mean values $\bar{X}_1 = 25$ and $\bar{X}_2 = 30$, sample standard deviations $s_1 = 3$ and $s_2 = 5$.

The relevant test statistic is given from earlier as a comparison of sample means

$$t = \frac{(\bar{X}_1 - \bar{X}_2) - (\mu_1 - \mu_2)}{s\sqrt{1/n_1 + 1/n_2}} \quad \text{where} \quad s^2 = \frac{(n_1 - 1)s_1^2 + (n_2 - 1)s_2^2}{n_1 + n_2 - 2}$$

and $\quad t \sim t_{[v]} \quad$ with $\quad v = n_1 + n_2 - 2$.

In the above μ_1 and μ_2 are the relevant population mean values which are unknown. However, an astute choice of hypothesis makes this unnecessary; the two hypotheses chosen are:

H_0: null hypothesis $\mu_1 = \mu_2$ (this will prove specific enough)
H_1: alternative hypothesis $\mu_1 \neq \mu_2$ (this defines a two-tailed test).

Assuming the null hypothesis applies then the test statistic becomes after evaluation:

$$t = \frac{(\bar{X}_1 - \bar{X}_2)}{s\sqrt{1/3}}$$

where $s = 4.123$ and $t \sim t_{[10]}$.

The data value for the test statistic is

$$\hat{t} = \frac{25 - 30}{4.123 \times 0.577} = -2.102$$

Taking a significance level of 0.05, then critical values are associated with two tails, each with a probability of 0.025 and upper value can be determined from Table 10.2 for $t_{0.025;10}$ as $c_2 = 2.23$. Thus it follows that the acceptance region for the null hypothesis is $-2.23 < t < 2.23$. The calculated data value lies within this region, from which it is possible to conclude that the null hypothesis is consistent with the data and therefore there is no significant difference between the strengths of the two mixes.

The test statistic specifically used in this section is often termed a t-statistic and can be usefully identified as

$$t = \frac{\text{observed difference in means}}{\hat{\sigma}_e}$$

10.5.3 Comparison of variances

Use can be made of the F-distribution to determine if any changes to a process has resulted in a smaller variation (measured by sample variances) rather than just to random choice as identified in the following example.

Example
The sample variances for the diameters of 23 nominally identical cast cylinders was 1.93 mm^2. For a random sample of size 13 taken from a second population the corresponding figure was 4.06 mm^2. Would one be justified in assuming that the two populations have diameters with the same variability?

The variates in this case are X_1 and X_2 the cylinder diameters from the two sources and the test is based on a comparison of population variances σ_1^2 and σ_2^2. A test statistic is given by

$$F = \frac{\text{largest sample value of } s_1^2}{\text{smallest sample value of } s_2^2}$$

and distributed as $F_{[v_1, v_2]}$, where $v_1 = n_1 - 1$, $v_2 = n_2 - 1$.

Comparing with the data values then $s_1^2 = 4.06$ (largest value), $n_1 = 13$ and $s_2^2 = 1.93$, $n_2 = 23$.

The relevant two hypotheses are:

H_0: null hypothesis $\sigma_1^2 = \sigma_2^2$ (i.e. $\sigma_1^2 / \sigma_2^2 = 1$)

H_1: alternative hypothesis $\sigma_1^2 \neq \sigma_2^2$ (this defines a two-tailed test).

The data value for the test statistic is evaluated as $\hat{F} = 4.06/1.93 = 2.10$.

Taking a significance level of 0.05, then critical values are associated with two tails, each with a probability of 0.025 but preliminary selection of taking the largest sample variance as the numerator means only the upper value is relevant. A critical upper value associated with a probability $\alpha = 0.025$ is given from a table value for $F_{0.025;12,22}$ and determined as $c_2 = 2.60$. Thus the upper rejection region is $\hat{F} > c_2$, and comparison of values gives that $\hat{F}$ lies below the rejection region and a conclusion is no significant difference between variation in the two populations.

The test statistic specifically used in this section is termed an F-statistic and can be expressed as

$$F = \frac{\sigma_1^2}{\sigma_2^2} \quad [\text{where } \sigma_1 > \sigma_2]$$

10.5.4 Significance and errors

An obvious factor with hypothesis testing is the choice and meaning of the significance level. By nature of dealing with processes that involve variation then some error is always present but it is important to try to quantify any error and evaluate any subsequent consequences. The significance level, α say, is in fact a probability measure:

$$P(\text{rejecting } H_0 \text{ when it is true}) = \alpha$$

and is the chance that the test statistic value $\hat{Z}$ falling in the rejection region within a hypothesis test could have occurred as a rare event rather than an incorrect hypothesis H_0. Such an error is called a **Type 1** error and obviously should be made as small as possible.

However, decreasing α increases the chance of making a **Type 2** corresponding to failing to rejecting the hypothesis H_0 when it is false. The value for this error β is given by the probability

$$P(\text{accepting } H_0 \text{ when it is false}) = \beta.$$

In sampling, a decision has to be made for the value of α and/or β by taking into account the costs and penalties attached to both types of errors. However, as might be expected these values are linked; generally decreasing Type 1 errors will increase Type 2 errors and vice versa. Calculation of the links between the parameters α, β and n depend upon the specific test chosen and can be very involved. In practice a decision rule for a significance test is determined by taking a significance level α and arrive at an appropriate rule; the associated values of β can be computed for various values of n. Alternatives that may be used are:

- given a decision rule – compute the errors α and β
- decide on α and β and then arrive at a decision rule.

A related curve is that of β known as the operating characteristic curve, or O-C curve, and is described in Chapter 9.

10.6 Regression models

10.6.1 Correlation

Correlation is concerned with the amount of association between two or more sets of variables. An illustration is given earlier in *Case 2* for the deflection of a concrete beam where the scattergram of the data displayed in Figure 10.2 shows a strong linear relationship, with increasing values of beam deflection (y) associated with proportionate increasing values of applied load (x). The amount of association for different situations may not be so discernible as illustrated in the following sets of data:

Case 3 An experimental determination of the relation between the normal stress (x) and the shear resistance (y) of a cement-stabilized soil yielded the following results:

Normal stress x kN/mm^2	10	12	14	16	18	20
Shear resistance y kN/mm^2	10.9	18.7	15.4	25.1	19.3	17.6

Case 4 Percentages of sand (y) recorded at different depths (x) from samples were:

x (mm)	0	400	800	1200	1600	2000	2400	2800	3200
y (%)	70.2	52.9	54.2	52.4	47.4	49.1	30.7	36.8	37.4

Case 5 Failure loads concrete beams with load for first crack (x) and failure load (y) were:

load x	10350	8450	7200	5100	6500	10600	6000	6000
load y	10350	9300	9600	10300	9400	10600	10100	9900
load x	9500	6500	9300	6000	6000	5800	6500	
load y	9500	10200	9300	9550	9550	10500	10200	

The relationships between values y and x can be individually plotted on a series of scattergrams as shown in Figure 10.12 from which the qualitative association between the variables x and their linked values y can be assessed.

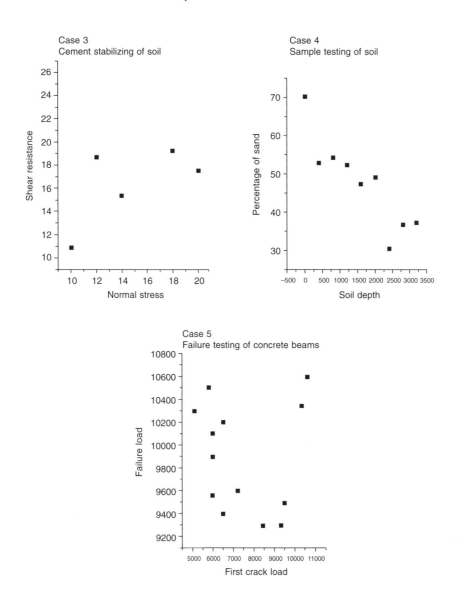

Figure 10.12 Scattergrams of data for Cases 3, 4 and 5.

In *Case* 4 then a trend exists where decreasing values of y are linked to increasing values of x; the data perhaps suggest a linear relationship could exist. Looking at the scattergram for *Case 5* it is difficult to perceive that any meaningful relationship exists between the variables x and y. Data from *Case 3* shows some trend exists in that increasing values of x are associated with increasing values of y, but the relationship is not obviously linear. Clearly, it would not make sense to seek a linear relationship for *Case 5* but that a linear relationship could reasonably be sought for *Cases 2* and *4*. For *Case 3* perhaps some tentative linear relationship could be determined, but some measure of confidence of how this relationship matches with the data values would be desirable. The measure of the association of variables is called correlation. The most widely used association relationship is that of a straight-line (linear) fit to data pair observations between a response variable (y) and an explanatory variable (x). In this section a linear fit approach will be assumed but in practice, care should be exercised to consider the possibility that some other form of relationship might be more appropriate.

10.6.2 Regression – least-squares method

Regression is a general term used in data analysis to mean 'trend' or 'pattern'. Many engineering problems are concerned with determining a relationship between a set of variables. Even for a strongly linear relationship, such as in *Case 2*, in practice all data points are unlikely to align due to random error. Generally, the data points are more scattered and a more realistic aim would be to obtain a 'best' curve through the collection of all data points and the most used method is to use a least-squares method.

Given a set of data $(x_1, y_1), \ldots (x_i, y_i) \ldots (x_n, y_n)$ consider fitting a straight line $y = a + bx$ through the data points so as to achieve some form of 'best fit'. In practice, if data values are plotted as illustrated in Figure 10.13 this means adjusting the slope of the line (parameter b) and the intercept on the y-axis (parameter a) until some form of optimal fit is achieved.

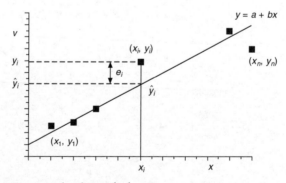

Figure 10.13 Least squares approximation method.

Calculated values for these parameters can be determined provided a measure of fit is defined. At any data value x_i, the data value is specified as y_i and the corresponding regression curve value is $\hat{y}_i = a + bx_i$; this defines an associated error $e_i = y_i - \hat{y}_i = y_i - a - bx_i$.

As the values of a and b are changed, then the error associated with each data point will change; 'best values' for a and b will occur when the errors e_i are minimized. A measure of the global error is given by the sum of squares

$$S = \sum_{i=1}^{n} e_i^2 = \sum_{i=1}^{n} (y_i - a - bx_i)^2$$

$S = S(a, b)$ is a function of the two variables a and b, and can be chosen such that S is minimized.

Applying appropriate calculus for a function of several variables, the result is:

$$b = \frac{n \sum x_i y_i - \sum x_i \sum y_i}{n \sum x_i^2 - (\sum x_i)^2}, \quad a = \bar{y} - b\bar{x}$$

where

$$\bar{x} = \frac{1}{n} \sum x_i \quad \text{and} \quad \bar{y} = \frac{1}{n} \sum y_i$$

These formulae determine the parameters associated with a least squares regression line as illustrated in a later example.

10.6.3 Correlation coefficient

A measure of the correlation between data values and a linear fit can be obtained as discussed below, and illustrated in Figure 10.14. At any data value x_i, then both a recorded data value y_i and a value $\hat{y}_i = a + bx_i$ calculated from the regression formula are available. An assessment of the closeness of agreement between the data values is obtained from considering the variations $(y_i - \hat{y}_i)$ between the recorded data value and the regression value at a general data point (x_i, y_i).

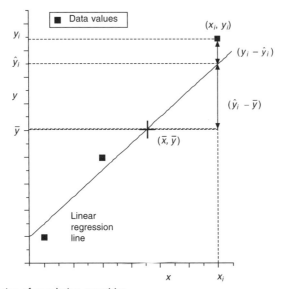

Figure 10.14 Calculation of correlation quantities.

The least-squares method can be shown to have the following properties that are useful in the following analysis:

(i) The regression line will pass through the point with coordinate $(\bar{x}, \bar{y})$, i.e. pass through the sample mean values of both x_i and y_i;

(ii) $\sum_{1}^{n} (y_i - \hat{y}_i)(\hat{y}_i - \bar{y}) = 0$

It is convenient to use the point $(\bar{x}, \bar{y})$ as a reference point, which from property 1 lies on the regression line. Any data value y_i can then be readily expressed as

$$(y_i - \bar{y}) = (y_i - \hat{y}_i) + (\hat{y}_i - \bar{y})$$

as illustrated in Figure 10.14.

The above expression is valid for all data points and squaring and summing over all n data values, and using property 2 gives a useful expression

$$\underbrace{\sum_{i=1}^{n} (y_i - \bar{y})^2}_{\text{total variation}} = \underbrace{\sum_{i=1}^{n} (y_i - \hat{y}_i)^2}_{\text{unexplained variation}} + \underbrace{\sum_{i=1}^{n} (\hat{y}_i - \bar{y})^2}_{\text{explained variation}}$$

The term $\sum_{i=1}^{n} (\hat{y}_i - \bar{y})^2$ is called the explained variation and corresponds to a measure if all data values were to lie exactly on the regression line. Discrepancies and variations from data points not lying on the regression line are given by the measure identified as the unexplained variation, and corresponding to the sum $\sum_{i=1}^{n} (\hat{y}_i - \bar{y})^2$. Clearly the requirement for good correlation is that the unexplained variation is relatively small, or equally, the total variation and the explained variation are almost equal. A useful relative measure of correlation is therefore given by

$$R^2 = \frac{\text{explained variation}}{\text{total variation}} = \frac{\sum (\hat{y}_i - \bar{y})^2}{\sum (y_i - \bar{y})^2} = \frac{SSR}{CSS}$$

$$= -\frac{\sum (y_i - \hat{y}_i)^2}{\sum (y_i - \bar{y})^2} = 1 - \frac{SSE}{CSS}$$

In the above, a number of quantities are determined directly from data values as:

$CSS = \sum_{i=1}^{n} (y_i - \bar{y})^2$ – computed sum of squares

$SSR = \sum_{i=1}^{n} (\hat{y}_i - \bar{y})^2$ – sum of squares due to regression

$SSE = \sum_{i=1}^{n} (y_i - \hat{y}_i)^2$ – sum of squares due to errors.

The quantity R^2 is called the coefficient of determination and is a useful measure of the association between a linear regression line and the data with $R^2 = 1$ corresponding to a perfect fit and $R^2 = 0$ corresponds to no dependence between x and y.

Values for the Case examples following from fitting a least-squares fit can be readily calculated as:

Case 2: $R^2 = 0.95$ indicates a very strong correlation between data values;
Case 3: $R^2 = 0.26$ indicates that correlation between data values is weak;
Case 4: $R^2 = 0.79$ indicates a fair correlation between data values;
Case 5: $R^2 = 0.00004$ indicates little correlation exists;

10.6.4 Example – Case 2 Beam deflection

The ten data values associated with *Case 2* are used to provide a regression calculation to obtain a least-squares fit, calculation of the correlation coefficient and analysis of residuals. It is convenient to display derived values in a spreadsheet format as follows:

i	x_i	y_i	$x_i - \bar{x}$	$(y_i - \bar{y})$	$\hat{y}_i = a + bx_i$	Residual $y_i - \hat{y}_i$
1	100	45	−45	−21.5	43.56	1.44
2	110	52	−35	−14.5	48.66	3.34
3	120	54	−25	−12.5	53.76	0.24
4	130	54	−15	−12.5	58.85	−4.85
5	140	62	−5	−4.5	63.95	−1.95
6	150	68	5	−1.5	69.05	−1.05
7	160	75	15	8.5	74.15	0.85
8	170	75	25	8.5	79.24	−4.24
9	180	92	35	25.5	84.34	7.66
10	190	88	45	21.5	89.44	−1.44

Derived values associated with the operational formulae used are:

$$\bar{x} = \frac{1}{10} \sum_{i=1}^{10} x_i = 145, \quad \bar{y} = \frac{1}{10} \sum_{i=1}^{10} y_i = 66.5$$

$$\sum_{i=1}^{10} (x_i - \bar{x})(y_i - \bar{y}) = 4205, \quad \sum_{i=1}^{10} (x_i - \bar{x})^2 = 8250, \quad \sum_{i=1}^{10} (y_i - \bar{y})^2 = 2264.5$$

Using the given formulae, $b = 4205/8250 = 0.510$ and $a = 66.5 - 0.51*145 = -7.41$.

Hence the least squares linear regression is $y = -7.41 + 0.51x$. Deflection values $\hat{y}_i$ obtained from the least-squares analysis are calculated in the spreadsheet together with residual values $(y_i - \hat{y}_i)$ for information; data values together with the regression curve are displayed in Figure 10.15.

Corresponding values for correlation quantities are:

CSS = 2264.5, SSR = 2143.3, SSE = 121.2, giving a value for the correlation coefficient $R^2 = 0.946$.

10.6.5 Analysis of residuals

The starting point for measuring regression error were the quantities $e_i = y_i - \hat{y}_i$, which are called the residuals. These may be usefully considered following a regression analysis

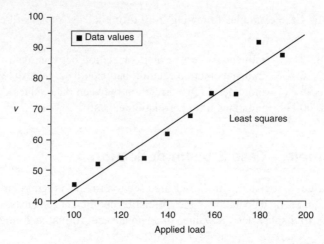

Figure 10.15 Comparison of least-squares linear regression with data values in Case 2 – Beam deflection.

to look at the level of agreement. For example, plotting the residuals for *Case 2* is shown in Figure 10.16. This shows residuals apparently randomly distributed about the zero error line; this is typical of data that is consistent with a linear regression.

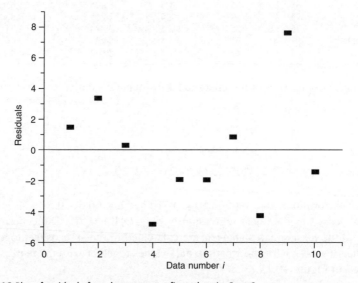

Figure 10.16 Plot of residuals from least-squares fit to data in Case 2.

The value of providing a plot of residuals is to identify if the choice of a linear fit (or other) is appropriate. If residuals follow a distinct trend, then a linear regression curve may not be appropriate, and a quadratic or other curve may be more appropriate. The residual curve can also highlight suspect data points (outliers), i.e. values that are distinct from the rest of the data and may arise from an uncharacteristic operational or measuring error. These outliers may often need special attention to decide whether they should be included within the analysis.

10.6.6 Extension to multivariate

In many practical cases, a dependant variable y may depend upon more than one independent variable, i.e. $y = f(x_1, x_2, \ldots x_m)$. A direct extension to the regression analysis is available to obtain a least squares linear fit of the form $y = \alpha_0 + \alpha_1 x_1 + \ldots \alpha_m x_m$ to the data.

In this case, analysis is best carried out using matrix algebra. Also as drawing a scattergram is not feasible, analysis of the level of fit is conducted through looking at the residuals and the values of the appropriate correlation coefficient. Details will not be covered within this chapter.

10.6.7 Fit of regression curves and confidence lines

Hypothesis testing techniques can also be used to test if a linear regression obtained to a set of data points is acceptable to a given significance. We have already seen that if R^2, the coefficient of determination is close to a value of 1, then a close fit (i.e. good correlation) is expected.

It can be shown that the quantity

$$F = \frac{R^2/v_1}{(1-R^2)/v_2}, \quad v_1 = m \quad \text{and} \quad v_2 = n - (m+1)$$

is distributed as an F-variate, $F_{[v_1,v_2]}$ where n is the number of data points and $m = 2$ for a linear fit.

Estimation techniques can be applied to linear regression analysis using the t-distribution with respect to confidence limits to the two parameters a and b for a linear fit. This gives rise to associated confidence limit values for each data value and joining these values produces confidence lines.

10.7 Statistical formulae and tables

Selected statistical formulae:

$$\sigma_{\bar{X}} = \frac{\sigma_X}{\sqrt{n}}$$

$$\sigma_{\bar{X}_1 - \bar{X}_2} = \sqrt{\frac{\sigma_1^2}{n_1} + \frac{\sigma_2^2}{n_2}}$$

$$\hat{\sigma}_e = s_p \sqrt{\frac{1}{n_1} + \frac{1}{n_2}} \quad \text{where} \quad s_p^2 = \frac{(n_1 - 1)s_1^2 + (n_2 - 1)s_2^2}{n_1 + n_2 - 2}$$

$$t = \frac{\text{observed difference in means}}{\hat{\sigma}_e}$$

$$F = \frac{\sigma_1^2}{\sigma_2^2} \quad [\text{where } \sigma_1 > \sigma_2]$$

$$CSS = \sum_{i=1}^{n} (y_i - \bar{y})^2$$

$$SSE = \sum_{i=1}^{n} (y_i - \hat{y}_i)^2$$

$$b = \frac{n \sum x_i y_i - \sum x_i \sum y_i}{n \sum x_i^2 - (\sum x_i)^2}$$

$$a = \bar{y} - b\bar{x}$$

Statistical Table 10.1 The normal distribution

Z is the normalized variate N(0, 1)

$f(z)$ is the probability density function (pdf) $= \dfrac{1}{\sqrt{2\pi}} e^{-\frac{1}{2}z^2}$

$F(z)$ is the cumulative distribution function $= \dfrac{1}{\sqrt{2\pi}} \displaystyle\int_{-\infty}^{z} e^{-\frac{1}{2}z^2}\, dz$

Values for $z < 0$ are given by: $f(-z) = f(z)$ and $F(-z) = 1 - F(z)$

Table values given from the Excel function NORMDIS

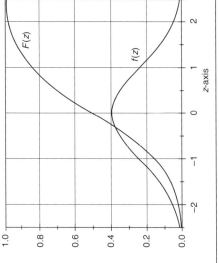

$f(Z)$	Z					$F(Z)$					
		0.00	0.01	0.02	0.03	0.04	0.05	0.06	0.07	0.08	0.09
0.3989	**0.0**	0.5000	0.5040	0.5080	0.5120	0.5160	0.5199	0.5239	0.5279	0.5319	0.5359
0.3970	**0.1**	0.5398	0.5438	0.5478	0.5517	0.5557	0.5596	0.5636	0.5675	0.5714	0.5753
0.3910	**0.2**	0.5793	0.5832	0.5871	0.5910	0.5948	0.5987	0.6026	0.6064	0.6103	0.6141
0.3814	**0.3**	0.6179	0.6217	0.6255	0.6293	0.6331	0.6368	0.6406	0.6443	0.6480	0.6517
0.3683	**0.4**	0.6554	0.6591	0.6628	0.6664	0.6700	0.6736	0.6772	0.6808	0.6844	0.6879
0.3521	**0.5**	0.6915	0.6950	0.6985	0.7019	0.7054	0.7088	0.7123	0.7157	0.7190	0.7224
0.3332	**0.6**	0.7257	0.7291	0.7324	0.7357	0.7389	0.7422	0.7454	0.7486	0.7517	0.7549
0.3123	**0.7**	0.7580	0.7611	0.7642	0.7673	0.7704	0.7734	0.7764	0.7794	0.7823	0.7852
0.2897	**0.8**	0.7881	0.7910	0.7939	0.7967	0.7995	0.8023	0.8051	0.8078	0.8106	0.8133
0.2661	**0.9**	0.8159	0.8186	0.8212	0.8238	0.8264	0.8289	0.8315	0.8340	0.8365	0.8389
0.2420	**1.0**	0.8413	0.8438	0.8461	0.8485	0.8508	0.8531	0.8554	0.8577	0.8599	0.8621
0.2179	**1.1**	0.8643	0.8665	0.8686	0.8708	0.8729	0.8749	0.8770	0.8790	0.8810	0.8830
0.1942	**1.2**	0.8849	0.8869	0.8888	0.8907	0.8925	0.8944	0.8962	0.8980	0.8997	0.9015
0.1714	**1.3**	0.9032	0.9049	0.9066	0.9082	0.9099	0.9115	0.9131	0.9147	0.9162	0.9177
0.1497	**1.4**	0.9192	0.9207	0.9222	0.9236	0.9251	0.9265	0.9279	0.9292	0.9306	0.9319

$f(Z)$	Z	$F(Z)$									
		0.00	0.01	0.02	0.03	0.04	0.05	0.06	0.07	0.08	0.09
0.1295	**1.5**	0.9332	0.9345	0.9357	0.9370	0.9382	0.9394	0.9406	0.9418	0.9429	0.9441
0.1109	**1.6**	0.9452	0.9463	0.9474	0.9484	0.9495	0.9505	0.9515	0.9525	0.9535	0.9545
0.0940	**1.7**	0.9554	0.9564	0.9573	0.9582	0.9591	0.9599	0.9608	0.9616	0.9625	0.9633
0.0790	**1.8**	0.9641	0.9649	0.9656	0.9664	0.9671	0.9678	0.9686	0.9693	0.9699	0.9706
0.0656	**1.9**	0.9713	0.9719	0.9726	0.9732	0.9738	0.9744	0.9750	0.9756	0.9761	0.9767
0.0540	**2.0**	0.9772	0.9778	0.9783	0.9788	0.9793	0.9798	0.9803	0.9808	0.9812	0.9817
0.0440	**2.1**	0.9821	0.9826	0.9830	0.9834	0.9838	0.9842	0.9846	0.9850	0.9854	0.9857
0.0355	**2.2**	0.9861	0.9864	0.9868	0.9871	0.9875	0.9878	0.9881	0.9884	0.9887	0.9890
0.0283	**2.3**	0.9893	0.9896	0.9898	0.9901	0.9904	0.9906	0.9909	0.9911	0.9913	0.9916
0.0224	**2.4**	0.9918	0.9920	0.9922	0.9925	0.9927	0.9929	0.9931	0.9932	0.9934	0.9936
0.0175	**2.5**	0.9938	0.9940	0.9941	0.9943	0.9945	0.9946	0.9948	0.9949	0.9951	0.9952
0.0136	**2.6**	0.9953	0.9955	0.9956	0.9957	0.9959	0.9960	0.9961	0.9962	0.9963	0.9964
0.0104	**2.7**	0.9965	0.9966	0.9967	0.9968	0.9969	0.9970	0.9971	0.9972	0.9973	0.9974
0.0079	**2.8**	0.9974	0.9975	0.9976	0.9977	0.9977	0.9978	0.9979	0.9979	0.9980	0.9981
0.0060	**2.9**	0.9981	0.9982	0.9982	0.9983	0.9984	0.9984	0.9985	0.9985	0.9986	0.9986

Statistical Table 10.2 Critical values of the *t*-distribution

The table gives the values of $t_{\alpha;v}$, the critical values for a significance level of α in the upper tail of the distribution. The *t*-distribution having v degrees of freedom.

Critical values for the lower tail is given by $-t_{\alpha;v}$ (symmetry).

For critical values of |*t*|, corresponding to a two-tailed region, the column headings for α must be doubled.

Table values given from the Excel function TINV

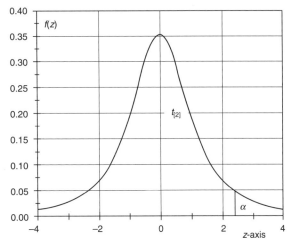

v \ $\alpha =$	0.1	0.05	0.025	0.01	0.005	0.001	0.0005
1	3.078	6.314	1.000	31.821	636.578	318.29	636.58
2	1.886	2.920	0.816	6.965	31.600	22.328	31.600
3	1.638	2.353	0.765	4.541	12.924	10.214	12.924
4	1.533	2.132	0.741	3.747	8.610	7.173	8.610
5	1.476	2.015	0.727	3.365	6.869	5.894	6.869
6	1.440	1.943	0.718	3.143	5.959	5.208	5.959
7	1.415	1.895	0.711	2.998	5.408	4.785	5.408
8	1.397	1.860	0.706	2.896	5.041	4.501	5.041
9	1.383	1.833	0.703	2.821	4.781	4.297	4.781
10	1.372	1.812	0.700	2.764	4.587	4.144	4.587
11	1.363	1.796	0.697	2.718	4.437	4.025	4.437
12	1.356	1.782	0.695	2.681	4.318	3.930	4.318
13	1.350	1.771	0.694	2.650	4.221	3.852	4.221
14	1.345	1.761	0.692	2.624	4.140	3.787	4.140
15	1.341	1.753	0.691	2.602	4.073	3.733	4.073
16	1.337	1.746	0.690	2.583	4.015	3.686	4.015
17	1.333	1.740	0.689	2.567	3.965	3.646	3.965
18	1.330	1.734	0.688	2.552	3.922	3.610	3.922
19	1.328	1.729	0.688	2.539	3.883	3.579	3.883
20	1.325	1.725	0.687	2.528	3.850	3.552	3.850
21	1.323	1.721	0.686	2.518	3.819	3.527	3.819
22	1.321	1.717	0.686	2.508	3.792	3.505	3.792
23	1.319	1.714	0.685	2.500	3.768	3.485	3.768
24	1.318	1.711	0.685	2.492	3.745	3.467	3.745
25	1.316	1.708	0.684	2.485	3.725	3.450	3.725
26	1.315	1.706	0.684	2.479	3.707	3.435	3.707
27	1.314	1.703	0.684	2.473	3.689	3.421	3.689
28	1.313	1.701	0.683	2.467	3.674	3.408	3.674
29	1.311	1.699	0.683	2.462	3.660	3.396	3.660
30	1.310	1.697	0.683	2.457	3.646	3.385	3.646
40	1.303	1.684	0.681	2.423	3.551	3.307	3.551
60	1.296	1.671	0.679	2.390	3.460	3.232	3.460
120	1.289	1.658	0.677	2.358	3.373	3.160	3.373
∞	1.282	1.645	0.675	2.327	3.291	3.091	3.291

Statistical Table 10.3 Critical values of the F-distribution

The table gives the values of $F_{\alpha;v_1,v_2}$, the critical values for a significance level of α in the upper tail of the distribution, the F-distribution having v_1 degrees of freedom in the numerator and v_2 degrees of freedom in the denominator.

For each pair of degrees of freedom of v_1 and v_2, the upper values $F_{\alpha;v_1,v_2}$ is tabulated for $\alpha = 0.05$ and 0.025 (bracketed).

Critical values corresponding to the lower tail of the distribution are obtained by the relation $F_{1-\alpha;v_1,v_2} = 1/F_{\alpha;v_2,v_1}$

Table values given from the Excel function FINV

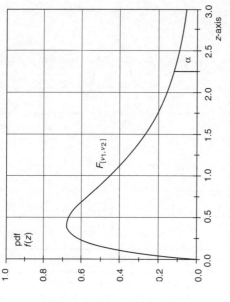

v_1 \ v_2	1	2	3	4	5	6	7	8	10	12	24	∞
1	161.4	199.5	215.7	224.6	230.2	234.0	236.8	238.9	241.9	243.9	249.1	254.3
	(648)	(799)	(864)	(900)	(922)	(937)	(948)	(957)	(969)	(977)	(997)	(1018)
2	18.5	19.0	19.2	19.2	19.3	19.3	19.4	19.4	19.4	19.4	19.5	19.5
	(38.5)	(39.0)	(39.2)	(39.2)	(39.3)	(39.3)	(39.4)	(39.4)	(39.4)	(39.4)	(39.5)	(39.5)
3	10.13	9.55	9.28	9.12	9.01	8.94	8.89	8.85	8.79	8.74	8.64	8.53
	(17.44)	(16.04)	(15.44)	(15.10)	(14.88)	(14.73)	(14.62)	(14.54)	(14.42)	(14.34)	(14.12)	(13.90)
4	7.71	6.94	6.59	6.39	6.26	6.16	6.09	6.04	5.96	5.91	5.77	5.63
	(12.22)	(10.65)	(9.98)	(9.60)	(9.36)	(9.20)	(9.07)	(8.98)	(8.84)	(8.75)	(8.51)	(8.26)
5	6.61	5.79	5.41	5.19	5.05	4.95	4.88	4.82	4.74	4.68	4.53	4.37
	(10.01)	(8.43)	(7.76)	(7.39)	(7.15)	(6.98)	(6.85)	(6.76)	(6.62)	(6.52)	(6.28)	(6.02)
6	5.99	5.14	4.76	4.53	4.39	4.28	4.21	4.15	4.06	4.00	3.84	3.67
	(8.81)	(7.26)	(6.60)	(6.23)	(5.99)	(5.82)	(5.70)	(5.60)	(5.46)	(5.37)	(5.12)	(4.85)
7	5.59	4.74	4.35	4.12	3.97	3.87	3.79	3.73	3.64	3.57	3.41	3.23
	(8.07)	(6.54)	(5.89)	(5.52)	(5.29)	(5.12)	(4.99)	(4.90)	(4.76)	(4.67)	(4.41)	(4.14)

v_2 \ v_1	1	2	3	4	5	6	7	8	10	12	24	∞
8	5.32 (7.57)	4.46 (6.06)	4.07 (5.42)	3.84 (5.05)	3.69 (4.82)	3.58 (4.65)	3.50 (4.53)	3.44 (4.43)	3.35 (4.30)	3.28 (4.20)	3.12 (3.95)	2.93 (3.67)
9	5.12 (7.21)	4.26 (5.71)	3.86 (5.08)	3.63 (4.72)	3.48 (4.48)	3.37 (4.32)	3.29 (4.20)	3.23 (4.10)	3.14 (3.96)	3.07 (3.87)	2.90 (3.61)	2.71 (3.33)
10	4.96 (6.94)	4.10 (5.46)	3.71 (4.83)	3.48 (4.47)	3.33 (4.24)	3.22 (4.07)	3.14 (3.95)	3.07 (3.85)	2.98 (3.72)	2.91 (3.62)	2.74 (3.37)	2.54 (3.08)
11	4.84 (6.72)	3.98 (5.26)	3.59 (4.63)	3.36 (4.28)	3.20 (4.04)	3.09 (3.88)	3.01 (3.76)	2.95 (3.66)	2.85 (3.53)	2.79 (3.43)	2.61 (3.17)	2.40 (2.88)
12	4.75 (6.55)	3.89 (5.10)	3.49 (4.47)	3.26 (4.12)	3.11 (3.89)	3.00 (3.73)	2.91 (3.61)	2.85 (3.51)	2.75 (3.37)	2.69 (3.28)	2.51 (3.02)	2.30 (2.72)
13	4.67 (6.41)	3.81 (4.97)	3.41 (4.35)	3.18 (4.00)	3.03 (3.77)	2.92 (3.60)	2.83 (3.48)	2.77 (3.39)	2.67 (3.25)	2.60 (3.15)	2.42 (2.89)	2.21 (2.60)
14	4.60 (6.30)	3.74 (4.86)	3.34 (4.24)	3.11 (3.89)	2.96 (3.66)	2.85 (3.50)	2.76 (3.38)	2.70 (3.29)	2.60 (3.15)	2.53 (3.05)	2.35 (2.79)	2.13 (2.49)
16	4.49 (6.12)	3.63 (4.69)	3.24 (4.08)	3.01 (3.73)	2.85 (3.50)	2.74 (3.34)	2.66 (3.22)	2.59 (3.12)	2.49 (2.99)	2.42 (2.89)	2.24 (2.63)	2.01 (2.32)
18	4.41 (5.98)	3.55 (4.56)	3.16 (3.95)	2.93 (3.61)	2.77 (3.38)	2.66 (3.22)	2.58 (3.10)	2.51 (3.01)	2.41 (2.87)	2.34 (2.77)	2.15 (2.50)	1.92 (2.19)
20	4.35 (5.87)	3.49 (4.46)	3.10 (3.86)	2.87 (3.51)	2.71 (3.29)	2.60 (3.13)	2.51 (3.01)	2.45 (2.91)	2.35 (2.77)	2.28 (2.68)	2.08 (2.41)	1.84 (2.09)
22	4.30 (5.79)	3.44 (4.38)	3.05 (3.78)	2.82 (3.44)	2.66 (3.22)	2.55 (3.05)	2.46 (2.93)	2.40 (2.84)	2.30 (2.70)	2.23 (2.60)	2.03 (2.33)	1.78 (2.00)
24	4.26 (5.72)	3.40 (4.32)	3.01 (3.72)	2.78 (3.38)	2.62 (3.15)	2.51 (2.99)	2.42 (2.87)	2.36 (2.78)	2.25 (2.64)	2.18 (2.54)	1.98 (2.27)	1.73 (1.94)
26	4.23 (5.66)	3.37 (4.27)	2.98 (3.67)	2.74 (3.33)	2.59 (3.10)	2.47 (2.94)	2.39 (2.82)	2.32 (2.73)	2.22 (2.59)	2.15 (2.49)	1.95 (2.22)	1.69 (1.88)
28	4.20 (5.61)	3.34 (4.22)	2.95 (3.63)	2.71 (3.29)	2.56 (3.06)	2.45 (2.90)	2.36 (2.78)	2.29 (2.69)	2.19 (2.55)	2.12 (2.45)	1.91 (2.17)	1.65 (1.83)
30	4.17 (5.57)	3.32 (4.18)	2.92 (3.59)	2.69 (3.25)	2.53 (3.03)	2.42 (2.87)	2.33 (2.75)	2.27 (2.65)	2.16 (2.51)	2.09 (2.41)	1.89 (2.14)	1.62 (1.79)
40	4.08 (5.42)	3.23 (4.05)	2.84 (3.46)	2.61 (3.13)	2.45 (2.90)	2.34 (2.74)	2.25 (2.62)	2.18 (2.53)	2.08 (2.39)	2.00 (2.29)	1.79 (2.01)	1.51 (1.64)
60	4.00 (5.29)	3.15 (3.93)	2.76 (3.34)	2.53 (3.01)	2.37 (2.79)	2.25 (2.63)	2.17 (2.51)	2.10 (2.41)	1.99 (2.27)	1.92 (2.17)	1.70 (1.88)	1.39 (1.48)
120	3.92 (5.15)	3.07 (3.80)	2.68 (3.23)	2.45 (2.89)	2.29 (2.67)	2.18 (2.52)	2.09 (2.39)	2.02 (2.30)	1.91 (2.16)	1.83 (2.05)	1.61 (1.76)	1.25 (1.31)
∞	3.84 (5.02)	3.00 (3.69)	2.60 (3.12)	2.37 (2.79)	2.21 (2.57)	2.10 (2.41)	2.01 (2.29)	1.94 (2.19)	1.83 (2.05)	1.75 (1.94)	1.52 (1.64)	1.00 (1.00)

Further reading

Benjamin, J.R. and Cornell C.A. *Probability, Statistics and Decision for Civil Engineers*. McGraw-Hill, New York.
Bowker, A.H. and Lieberman G.J. *Engineering Statistics*. Prentice Hall, Engleword Cliffs, NJ.
Chatfield, C. *Statistics for Technology*. Penguin Books, London.
Kennedy, J.B. and Neville, A.M. *Basic Statistical Methods for Engineers and Scientists,* 3rd edn. Harper & Row, New York.
Metcalfe, A.V. *Statistics in Engineering – A practical approach*. Chapman and Hall, London.
Metcalfe, A.V. *Statistics in Civil Engineering*. Arnold, London.
Miller, I. and Freund, J.E. *Probability and Statistics for Engineers*, 2nd edn. Prentice Hall, Englewood Cliffs, NJ.
Ross, S.M. *Introduction to Probability and Statistics for Engineers and Scientists*. Wiley, Chichester.
Smith, G.N. *Probability and Statistics in Civil Engineering – An Introduction*. Collins, London.
Walpole, R.E. and Myers, R.H. *Probability and Statistics for Engineers and Scientists*, 4th edn. Macmillan, London.

11

Standards, specifications and codes of practice

Bryan K. Marsh

11.1 Aims and objectives

The intention of this chapter is to give an introduction to standards, codes of practice and specifications. Particular emphasis is placed on their coverage of durability and how that is related to research and feedback from practice. It explains the differences between the three main types of document, their purpose and their status, and gives some background into their development and updating. Some general rules are given for specification writers and users.

11.2 Introduction

Standards, specifications and codes of practice are probably not the most interesting subject for a book on advanced concrete technology. Nevertheless, they are an essential feature of most concrete technologists' lives, except perhaps the most esoteric of laboratory-based researchers. While probably not succeeding in making the subject interesting, this chapter will attempt to unravel some of the mysteries and, perhaps, make them a little less foreboding.

11.3 Standards, specifications and codes of practice

The terms 'standard', 'specification' and 'code of practice' are used slightly differently in different parts of the world so it is only possible to give a few general rules about the coverage and authority of each of these types of document. It is, therefore, important to understand the particular hierarchy and authority of the various documents applicable in the place where one is working.

11.3.1 Standards

The International Standards Organization, (ISO, www.iso.ch) defines standards as:

> ... documented agreements containing technical specifications or other precise criteria to be used consistently as guidelines, or definitions of characteristics, to ensure that materials, products, processes and services are fit for their purpose.

Standards thus generally contain *requirements* that must be enforced if compliance with that standard is required or claimed. They commonly cover either products or test methods:

- *Product standards* exist to give specifiers and users confidence in the properties, performance and fitness for purpose of a particular type of product.
- *Test method standards* are intended to ensure measured properties are obtained by reliable and reproducible methods and that test results can be widely understood and compared.

11.3.2 Codes of practice

Codes of practice, or 'codes' as they are often termed, generally contain *recommendations* for design and/or construction. Many codes of practice are produced by national standards bodies (e.g. British Standards Institution) or other major national bodies (e.g. American Concrete Institute). Other codes may be issued by interested parties and may therefore have lower status, e.g. 'Code of practice for the safe erection of precast concrete cladding' (Architectural Cladding Association, 1998).

11.3.3 Specifications

Specifications are the specifier's means of conveying particular requirements to the constructor, supplier or producer. They may be standard specifications (e.g. local or national highways authorities, national airports authorities) or individually tailored to a particular job or to particular circumstances. Standard specifications may, in some applications and in some countries, contain similar information to codes of practice in other countries. For example, the Japanese Architectural Standard Specification, Chapter 5 (JASS-5, 1997, Reinforced concrete works) and the American specification for structural concrete (ACI 301-99) cover similar ground to the British code of practice for structural use of concrete (BS 8110-1, 1997). For the purposes of this book it is convenient to distinguish between standard specifications and project- or company-specific specifications. The latter two types are covered separately at the end of this chapter.

11.3.4 Other relevant documents

Although strictly outside the scope of this chapter there are many other important documents that provide useful information to back up standards, specifications and codes of practice. These include:

- Informative documents, such as product data sheets, Agrément certificates, etc.
- Opinion-forming documents, such as research papers, state-of-art reports, etc.
- Guidance documents, such as Concrete Society Technical Reports, CIRIA reports, RILEM recommendations, ACI Committee Reports, etc.

11.3.5 The role and status of standards, standard specifications and codes of practice

These documents exist to give guidance on a common minimum standard established as a result of research and experience. They are usually produced by committees of representatives of all appropriate aspects of the industry.

In most countries, they are not, in themselves, intended to have legal status neither are they intended as textbooks of current practice. Nevertheless, they are reviewed regularly and updated as necessary to reflect changes in knowledge, materials and practice. It is important that they represent a consensus view and it may thus take some time for the latest research findings to gain sufficiently wide acceptance to justify their inclusion. In this sense they tend to follow established good practice rather than to lead the way. Drafts of new national standards and technical amendments to existing ones are often sent out for public comment prior to their issue. Comments received at this stage are taken very seriously and reviewed in detail before agreeing on the final content.

It is important to distinguish between the use of the words 'should' and 'shall' in certain standards (e.g. BSI British standards, CEN European standards, ACI American standards). Standards use 'shall' to indicate where something is a *requirement*. Standards also use 'should' where an item is only a *recommendation* or guidance. Some nations' standards do not make such a distinction.

Standards may have a different legal status in different countries. For example, standards in Germany and Russia are part of the law and therefore compliance is obligatory. In many other countries such as the UK, standards are not part of the legal framework except where they are called up as part of a contract, in which case they assume the same legal status as the contract itself.

The consideration of durability or structural integrity, for example, is not simply a matter of satisfying the requirements or recommendations of a standard, code of practice or standard specification. As stated above, the standard or code should be taken only as providing a means of meeting a common minimum standard and this may be insufficient for some applications where, for example, the acceptable risk of failure is particularly small. Indeed, the foreword to every British Standard contains an important statement:

Compliance with a British Standard does not of itself confer immunity from legal obligations.

The American building code (ACI 318-02) makes its status clear through the following statements in the introductory section:

A building code states only the minimum requirements necessary to provide for public health and safety. The code is based on this principle. For any structure, the owner or structural designer may require the quality of materials and construction to be higher than the minimum requirements necessary to protect the public as stated in the code. However, lower standards are not permitted.

The code has no legal status unless it is adopted by the government bodies having the police power to regulate building design and construction. Where the code has not been adopted, it may serve as a reference to good practice even though it has no legal status.

In many countries, it is quite legal to work outside the requirements or recommendations of a standard although there is obviously a much greater risk to be borne by the engineer in ensuring the durability and structural integrity of the construction. An example of this in the UK is the use of concrete containing silica fume as part of the cementitious component. Such concrete is known to have many beneficial properties but silica fume is not, as yet, recognized within the UK concrete standard (BS 5328-1, 1997), as a cementitious component in concrete. Replacement of BS 5328 by the British version of the European standard for concrete (BS EN 206-1, 2000) in December 2003 will, however, remedy this situation. Similar situations will, no doubt, continue to arise as new materials, construction methods and design techniques are developed.

Where product standards exist, it is essential that a product supplier or manufacturer complies with at least this minimum standard. This also applies to component materials used in the manufacture of products (e.g. cement, aggregates, additions and admixtures in readymixed concrete).

11.3.6 Selection of appropriate standards and codes of practice

It is important that the most appropriate and up-to-date standard is selected for the task being undertaken to ensure the correct guidance is obtained. An engineer designing an element of a bridge, with a long design life (say, 120 years) and severe exposure conditions (de-icing salts, in particular), but using a general code of practice for structural concrete, more applicable to building structures, is likely to underestimate the durability requirements, unless the code specifically addresses those conditions. In such circumstances, the engineer could be held to be negligent.

Useful sources of information on available standards and codes of practice and their current status include:

- Websites of national standards bodies which sometimes contain lists of all their national standards

- The BSI Catalogue, in paper or CD form, contains information on all current British Standards including CEN and ISO standards published in the UK
- Collated volumes of ASTM (American) test method and product standards, by subject, which are published yearly
- *ACI Manual of Concrete Practice* in which ACI (American Concrete Institute) standards, standard specifications and other authoritative documents such as guides and state-of-art reports are published yearly

Hard copies of standards, standard specifications and codes of practice should be controlled documents within a Quality Management System to ensure they are always up to date. Copies of superseded and amended standards should be retained for future reference in case queries arise relating to their content when, for example, certain works were performed and it is essential to know what standards were in force at that time. These must, of course, be stored separately from current editions and clearly marked with their status to avoid accidental use.

The following are given as examples of coverage of various codes and the list is by no means comprehensive.

In the UK, BS 8110 (BS 8110-1, 1997) is concerned with the use of concrete in buildings and structures generally. More specific guidance for particular types of structures may be found in other British Standards for bridges (BS 5400-4, 1990), for maritime structures (BS 6349-1, 2000), for water-retaining structures (BS 8007, 1987), for low-rise buildings (BS 8103, 1995), and for agricultural buildings and structures (BS 5502, 1990). There are also various other standards for specific precast concrete elements.

In the USA, there is a general building code for structural concrete (ACI 318-02) but for specialized structures it will be necessary to refer to other documents such as the standard specification for cast-in-place architectural concrete (ACI 303.1-97) and the code requirements for nuclear safety-related concrete structures (ACI 349.1-91).

11.3.7 Worldwide use of standards

Most developed countries have their own national standards organization which produces the standards for use in that country. In some cases, however, the standards themselves may be based on those of other countries, or may just list variations from another country's standards, and give any extra requirements. For example, some Canadian standards are based on US standards and some Irish standards are based on British ones. It is also common to find British or American standards, for example, used in many other countries, particularly that may be or have been dependencies of those larger states. Caution should be exercised in the use of standards outside the geographic area in which they were intended for use, particularly with respect to the use in hot climates of British Standard provisions for durability, written for use in a temperate climate.

11.3.8 European Standards and International Standards

The need for harmonized or unified standards has never been greater and will only continue to rise as raw materials (such as cement and silica fume, in particular) increasingly become part of the import/export trade; as multinational companies supply products (such as admixtures, in particular) into many individual countries; and as concrete products are traded on international markets. It makes little sense therefore for the specification for Portland cement, water-reducing admixtures or precast concrete pipes, for example, to be different from one country to another where the job they are required to do is actually the same. With adequate allowance for variations due to specific local conditions, such as seismic conditions or the extremes of exposure environment, it also makes no sense for a designer working, say, in Ireland to have to use a completely different design code for a notionally identical structure than when working, say, in Russia.

Change is always met with resistance and, in practice, replacement of national standards by common standards may incur significant short-term costs but the long-term benefits to clients, specifiers, suppliers and constructors alike are not difficult to see to anyone facing in the right direction, i.e. forwards.

International Standards Organization (ISO)

In an ideal world standards, codes of practice and standard specifications would be universal but, as has been described earlier in this chapter, standards are usually created within individual countries. Nevertheless, the International Standards Organization (ISO) was set up in 1947 as a worldwide federation of national standards bodies from some 140 countries. It describes its mission thus:

> ... to promote the development of standardization and related activities in the world with a view to facilitating the international exchange of goods and services, and to developing co-operation in the spheres of intellectual, scientific, technological and economic activity.

It is probably safe to say, however, that the impact of ISO on standards for concrete and concrete construction is not large. Approximately 18 ISO standards exist for concrete-related test methods, and development work continues in areas such as performance requirements for concrete structures and a simplified design standard. Nevertheless, most countries prefer to use their own national standards or those from other sources but with high status such as US or British Standards. A major recent advance in harmonization of diverse national standards has, however, taken place within Europe under the guise of the European Standards Organization (CEN).

European Committee for Standardization (CEN)

Within Europe, eventually, the national Standards of all CEN members will be replaced by Eurocodes and other European Standards produced by CEN. Considerable progress has been made in the field of concrete technology and construction; indeed, this chapter is being written during a time of rapid transition from national to European Standards.

At the time of writing, the European Committee for Standardization (CEN) comprises the national standards organizations of:

Austria (ON), Belgium (IBN/BIN), Czech Republic (CSNI), Denmark (DS), Finland (SFS), France (AFNOR), Germany (DIN), Greece (ILOT), Hungary (MS 2T), Iceland (STRI), Ireland (NSAI), Italy (UNI), Luxembourg (SEE), Malta (MSA), The Netherlands (NNI), Norway (NSF), Portugal (IPQ), Slovakia (SUTN), Spain (AENOR), Sweden (SIS), Switzerland (SNV) and The United Kingdom (BSI).

The following Affiliates are expected to become full members in the coming years:

Albania (DPS), Bulgaria (SASM), Croatia (DZNM), Cyprus (CYS), Estonia (EVS), Latvia (LVS), Lithuania (LST), Poland (PKN), Romania (ASRO), Slovenia (SMIS) and Turkey (TSE).

European Standards and Eurocodes

When Eurocodes and European Standards come fully into force, any conflicting national standards or parts of standards are required to be withdrawn. European Standards are normally preceded by prestandards that have no legal status and exist alongside current national standards. These are generally issued by national standards bodies for trial use. Indeed, the draft Eurocode 2 for design of concrete structures (ENV 1992-1-1, 1992) was issued in 1992 and has been available for trial use since then. Many European Standards contain a national foreword and annex, or national application document, to assist with its use in individual countries.

In theory, prestandards have a maximum life of three years but this has been long exceeded in many cases. Each should then be either:

- converted to a full European Standard (EN)
- extended (once only) for a further two years
- replaced by a revised prestandard, or
- withdrawn.

The European Standard for concrete (EN 206-1, 2000) was issued in 2000 but implementation will be delayed until December 2003 when a package of essential supporting standards, including products and test methods, is completed. In some countries, a complementary Standard (e.g. BS 8500, 2002 in the UK) has been produced to accompany the European Standard to help in its use in a particular country.

European Standards and Eurocodes are issued by each of the CEN countries within the publication covers of its national standards body. Thus, for example, EN 206-1 is issued in Germany as a DIN standard, in France as an AFNOR standard, and in the UK as a British Standard (BS). The cover pages may contain a national foreword and, for Eurocodes, a national annex or national application document containing particular national rules necessary for implementation of the Eurocode.

Relevant CEN standards

Structural design, loading, fire design and general design for durability are contained in Eurocodes. The ones of particular relevance to concrete are Eurocode 1 (ENV 1991-1, 1996) which contains general rules, and the various parts of Eurocode 2 (ENV 1992-1-1, 1992) which cover design of concrete structures including buildings and bridges, etc. Another Eurocode will cover composite steel and concrete structures. At the time of writing these documents had not been published as full standards and are referred to here only in their prestandard (ENV) form.

Specification, production and conformity testing of concrete is contained in EN 206-1 (EN 206-1, 2000) which also contain the classification system for exposure environment. Prescriptive tables of limitations on mix composition for durability were intended but consensus could not be reached during development of the standard. Only a single table is contained in an informative annex and applies only to CEM I cement. Guidance applicable in the place of use of the concrete for composition limitations for durability will be contained in national documents such as BS 8500 (BS 8500, 2002) in the UK. This is not the favoured outcome of the CEN committee but it is likely that had they waited for consensus then the standard may never have seen light of day.

11.4 Prescription-based standards and performance-based standards

Standards, codes and specifications may take one of three approaches:

- Prescription-based
- Performance-based
- A combination of prescription and performance

11.4.1 The prescription-based approach

A prescription-based document will state requirements in terms of the means to be used to achieve the required performance. Thus a requirement to achieve freeze–thaw resistance may dictate:

- The type of cement, including limits on proportions of additions such as fly ash, blastfurnace slag or silica fume
- The amount of air to be entrained (specified as a minimum value, or a target value with tolerances)
- Minimum cement content
- Maximum free water–cement ratio
- Minimum compressive strength
- Limitations on aggregate (freeze–thaw resistant in accordance with a given test method)

The concrete supplier thus has little flexibility in the choice of concrete to be supplied. On the other hand, concrete supplied in accordance with such prescriptive requirements will be 'deemed-to-satisfy' in terms of freeze–thaw resistance. The actual performance will thus not be the responsibility of the concrete supplier.

Situations where prescription-based specification may be appropriate include:

- Where it is necessary to comply with particular requirements within a design code which is, in itself, prescriptive
- Where the specifier has a greater knowledge or experience in achieving the required goals than the concrete supplier
- Where experience has shown that this is the most appropriate approach for the particular circumstances

The success of prescription-based specification relies on the achievement of a reasonable level of workmanship as will have been assumed in development of the specification (e.g. compaction, curing, achievement of cover to reinforcement).

11.4.2 The performance-based approach

The concept of performance-based specification can be interpreted in at least two ways:

- Where the required performance is specified (with little or no limitation or guidance on how it should be achieved)
- Where the required performance is specified in terms of a test method and an acceptance level for the parameter being measured

Indeed, the concept of performance specification is difficult to apply as a single concept throughout the construction process for a given structure or element. For example, in the case of designing against chloride-induced corrosion of reinforcement, the *client* may have the clear performance requirement that the structure 'shall perform its required function for the [stated] required service life without the need for unplanned maintenance or repair'. The *designer* could then provide the constructor with a more detailed performance specification that, for example, 'the chloride level at the reinforcement, at a specified minimum cover to satisfy structural requirements, shall not exceed 0.8% by mass of cement within 60 years of commencement of use of the structure'. In turn, the *constructor* could (wisely or unwisely) translate this as a requirement to the concrete supplier that the concrete, of a given minimum strength level to satisfy structural requirements, shall have a chloride diffusion no greater than a certain value when tested under specified conditions in a given test method. Thus each level of performance specification allows the recipient of the specification flexibility in the approach to achieving the required performance.

Where just the required performance is specified it will be difficult to judge, at completion of construction, whether the required long-term performance will indeed be achieved in practice. This may have significant contractual implications. Where performance is specified in terms of an *in-situ* test method success may depend on the appropriateness of the selected test, and acceptance level, to the actual conditions existing in service. If the test is performed on test specimens then, just as for prescription-based specification, success may be dependent upon levels of workmanship.

Performance-based specification of concrete

The European Standard for concrete (EN 206-1, 2002) allows an essentially performance-based approach in the specification of designed concrete whereby the only information required to be specified comprises:

- Exposure class
- Compressive strength class (for structural purposes)
- Maximum nominal aggregate size
- Chloride content class (maximum permissible level of chloride in the concrete, as supplied)
- Consistence class

The national standards applicable in the place of use of the concrete contain the basic requirements (albeit in prescriptive terms) for concrete to meet the requirements of the various exposure classes but the concrete supplier is free to choose between the available options within these standards. In the future it may be possible to replace the current prescription-based requirements in the national standards by minimum levels of performance based on specified test methods.

11.5 The treatment of durability in standards, codes of practice and standard specifications

11.5.1 Introduction

Achievement of durability in a concrete structure is, of course, reliant on the satisfaction of many requirements throughout the construction process. It is thus hardly surprising that there is no single standard or code of practice on durability. The following section indicates briefly the contribution towards durability of various requirements in a large range of different standards, codes of practice and standard specifications. When used correctly these standards come together to provide the framework for provision of durable concrete structures.

11.5.2 Design for durability

Some structural design codes contain basic but essential guidelines on the various factors that need to be considered by the structural designer, and the ways in which these can influence durability. For example, the draft European Standard for the basis of design of structures (ENV 1991-1, 1996, but which will become EN 1990 'Basis of design' when completed and published) contains the following statement of the responsibilities of the designer:

> A structure shall be designed and executed in such a way that it will, during its intended life with appropriate degrees of reliability and in an economic way:
>
> – remain fit for the use for which it is required; and
> – sustain all actions and influences likely to occur during execution and use

It is intended by the standard that the expressions 'fit for the use' and 'actions and influences' should include durability in both cases. The standard continues . . .

> These requirements shall be met by the choice of suitable materials, by appropriate design and detailing, and by specifying control procedures for design, production, execution and use relevant to the particular project.
>
> It is an assumption in design that the durability of a structure or part of it in its environment is such that it remains fit for use during the design working life given appropriate maintenance.

An indication of required design working life from the same standard is given in Table 11.1

This highlights the inseparable influences of design, materials and execution, and the need to consider the required life of the structure and essential/likely maintenance.

The prestandard Eurocode 1 (ENV 1991-1, 1996) lists the following interrelated factors which, it requires, shall be considered to ensure an adequately durable structure:

Table 11.1 Required design working life classes from prestandard Eurocode 1 (ENV 1991-1, 1996)

Class	Required design working life (years)	Example
1	[1–5]	Temporary structures
2	[25]	Replaceable structural parts, e.g. gantry girders, bearings
3	[50]	Building structures and other common structures
4	[100]	Monumental building structures, bridges, and other civil engineering structures

- the intended and possible future use of the structure
- the required performance criteria
- the expected environmental influences
- the composition, properties and performance of the materials
- the choice of the structural system
- the shape of members and the structural detailing
- the quality of workmanship, and level of control
- the particular protective measures
- the maintenance during the intended life

Despite the deterioration of reinforced concrete being a time-dependent process, few design standards worldwide have traditionally included any indication of the life that may be expected under the various exposure conditions even when the requirements of that standard have been satisfied. Nevertheless, a new generation of standards for concrete is beginning to address this need and link their durability requirements with a specified design life (e.g. European Standards, JASS-5). This has, however, exposed limitations in the data available on which to base predictions of service life, or with which to design for a specific life. It can only be hoped that this development in standards will, for once, drive the need for more focused analysis of existing research information and feedback from service, if not actually generating new research. Such an exercise was performed in the UK (Hobbs, 1998) in the development of the new British Standard to complement the European Standard for concrete (BS 8500, 2002).

11.5.3 Constituent materials

General

In order that concrete is durable it is essential that each of its component parts is of adequate quality to achieve that goal. All the normal constituent materials for concrete are covered by standards that, provided the material is demonstrated to be in compliance, ensure a minimum acceptable level of quality for use in concrete. It would, however, be neither interesting nor particularly useful to include here an exhaustive list of these standards from the major nations. Indeed, the American standard specification for concrete (ACI 301-99) itself contains a list of some 81 American standards cited within that document alone. Nevertheless, it is, perhaps, useful to consider the major relevant features of these groups of standards in the next few sections.

Water

The American building code (ACI 318-99), for example, requires that mix water should be clean and free from injurious amounts of oils, acids, alkalis, salts, organic materials, or other substances deleterious to concrete or reinforcement. Non-potable water is permitted only if mortar tests show the strength, at 7 and 28 days, to be at least 90 per cent of identical specimens made with potable water. As well as requirements for strength some standards for mix water give limits on the effect on setting time, which can be adversely increased by some organic contaminants.

Cement

The number of different cements available worldwide is very large; indeed the European Standard for cement (EN 197-1, 2000) lists 27 different common cements including Portland cement and those containing additional materials such as fly ash, silica fume and blastfurnace slag. Nevertheless, most standards for cement contain a number of common requirements which ensure the cement is suitable for appropriate use in concrete. These include:

- Limits on chemical composition, to help ensure particular properties (e.g. a maximum limit on C_3A in some Portland cement to impart sulfate-resistance)
- Limits on fineness to ensure, among other things, that bleed behaviour is predictable
- Limits on setting time to ensure that finishing can be completed and strength gain is commenced within a reasonable time
- Limits on soundness, to ensure undesired expansion does not occur in use
- Minimum, and sometimes maximum, strength and rate of strength gain, to ensure adequate reactivity and structural performance, particularly at early ages.

A significant recent development is the issue of an ASTM standard for cement (ASTM C1157-00) which is essentially performance-based and does not contain fixed limits on composition. It classifies cement by type based on specific requirements for general use, high early strength, resistance to attack by sulfates, and heat of hydration.

Standards for cements for special purposes may have other requirements relating to those purposes. For example, low-heat cement may have a limit on heat of hydration.

Limitations on which cements are suitable in which applications are generally included in standards for concrete rather than those for the cements themselves. It should be unnecessary to state in a book such as this that the use of a quality cement does not, in itself, ensure durable concrete.

Additions

Using the new terminology of the European Standards (EN 206-1, 2000), additions are finely divided materials such as fly ash, ground granulated blastfurnace slag, silica fume, metakaolin, natural pozzolana and limestone powder. They may be reactive (type II) or nearly inert (type I). Common features of many standards for additions include limits on:

- Chemical composition
- Fineness, which is often related to reactivity
- Loss on ignition, to minimize carbon content which may have an adverse effect on air entrainment
- Reactivity, to ensure that type II additions provide the anticipated benefits.

Aggregates

Standards cover aggregates from natural sources, lightweight aggregates and manufactured aggregates. They are intended to ensure that the aggregates complying with them are suitable for appropriate use in concrete without fear of unexpected poor performance or premature deterioration, for example. Use of an aggregate complying with the appropriate standard cannot, however, on its own ensure satisfactory performance of the concrete.

Some concrete for special purposes may require aggregates with specific properties that are not covered by general aggregate standards and these will need to be considered within individual specifications. Examples might include limits on the 'polished stone value' and abrasion resistance for aggregates for use in concrete pavements.

Common features of many standards for aggregates include:

- Basic limits on composition, particularly to ensure the quantity of potentially deleterious material is within acceptable limits
- Particle shape, to ensure satisfactory workability and strength in concrete
- Grading, to ensure satisfactory cohesion and particle packing within concrete
- Soundness, to ensure the aggregate does not undergo excessive volume changes in concrete under service conditions such as freezing and thawing or wetting and drying.

Although it is usually desirable to employ aggregates that comply with relevant standards some concrete standards, such as the American building code (ACI 318-99), realize that this not always possible and permit exceptions for aggregates that have been proven by means of special testing or satisfactory performance experience of adequate strength and durability from actual service.

Admixtures

Standards for admixtures are more generally based on potential performance than on chemical composition. Nevertheless, they also often contain requirements relating to basic properties of concrete, such as setting time, compressive strength and length change, to ensure the admixture does not have a deleterious effect on the concrete in which they are used.

Common features of many standards for admixtures, such as the European Standard (EN 934-2, 2001), include:

- Basic limits on composition, such as chloride content, alkali content and pH.
- Performance requirements based on the primary function(s) of the admixture, such as freeze–thaw performance for air-entraining admixtures, and effect on setting time for retarders or accelerators. Water-reducing admixtures, for example, have to meet requirements for water reduction, and the effect on compressive strength, air content and bleeding. Air-entraining agents have to meet requirements for air content and air void characteristics in hardened concrete, and water-resisting admixtures have to meet minimum requirements for capillary absorption.

Reinforcement

Common features of many standards for reinforcement for concrete include:

- Chemical composition of the steel
- Bar sizes
- Yield stress, fatigue and rebending performance

Of particular interest, from the durability point of view, are the standards for stainless steel and, perhaps, epoxy-coated reinforcement. Stainless steel comes in many grades, not all of which offer adequate performance for sufficiently enhanced durability in concrete in severe exposure conditions compared to carbon steel reinforcement. The British Standard for stainless steel reinforcement (BS 6744, 2001) contains guidance on selection of the appropriate grade for corrosion resistance under different service conditions. On the other hand, the International Standard for epoxy-coated reinforcement (ISO 14654, 1999) contains a single coating thickness and no guidance on its suitability for use in aggressive service conditions.

11.5.4 Constituent materials test methods

In order that confidence can be expressed in the potential performance of the various constituents of concrete, it is essential that any tests that are performed to determine their properties are done so by reliable and standardized test methods. There is thus a very large number of test methods which have been standardized to support the standards dictating acceptable properties of the constituent materials. It is essential, for example, that the strength class of all cements within a given market is measured by the same method to enable direct comparison to be made. Unfortunately this is not always the case. For example, the standards for Portland cement in Europe, the USA and Russia all use the same specimen type and test method but the water–cement ratio of the tested mortar is 0.5, 0.485 and 0.4 respectively.

As mentioned previously, aggregates may be required to have specific properties depending on the particular application of the concrete in which they are used. Standards for aggregates thus require many standardized test methods including, for example:

- soundness
- aggregate abrasion value
- polished stone value
- aggregate crushing value
- drying shrinkage
- frost heave

Although the results from tests may not be directly applicable, even meaningful, to the practising engineer or concrete supplier, reassurance can be obtained by comparison of results with limiting values within standards or specifications. This is only possible if the tests are standardized.

11.5.5 Classification of exposure environment

If achievement of durability is not just to be left to chance then it is essential that the exposure environment to which concrete is to be subjected in service is correctly identified and characterized. As most designers are not experts in concrete technology, it is necessary to divide typical exposure environments into classes to enable recommendations to be given for means to provide adequate durability.

The exposure classification from the European Standard for concrete (EN 206-1, 2000) is given in Table 11.2.

Table 11.2 Exposure classification system from the European Standard for concrete (EN 206, 2000)

Class designation	Description of the environment	Informative examples where exposure classes may occur
1 No risk of corrosion or attack		
X0	For concrete without reinforcement or embedded metal: all exposures except where there is freeze–thaw, abrasion or chemical attack For concrete with reinforcement or embedded metal: very dry	Concrete inside buildings with very low air humidity

2 Corrosion induced by carbonation
Where concrete containing reinforcement or other embedded metal is exposed to air and moisture, the exposure shall be classified as follows:
Note: The moisture condition relates to that in the concrete cover to reinforcement or other embedded metal, but in many cases, conditions in the concrete cover can be taken as reflecting that in the surrounding environment. In these cases classification of the surrounding environment may be adequate. This may not be the case if there is a barrier between the concrete and its environment.

XC1	Dry or permanently wet	Concrete inside buildings with low air humidity Concrete permanently submerged in water
XC2	Wet, rarely dry	Concrete surfaces subject to long-term water contact Many foundations
XC3	Moderate humidity	Concrete inside buildings with moderate or high air humidity External concrete sheltered from rain
XC4	Cyclic wet and dry	Concrete surfaces subject to water contact, not within exposure XC2

3 Corrosion induced by chlorides other than from sea water
Where concrete containing reinforcement or other embedded metal is subject to contact with water containing chlorides, including de-icing salts, from sources other than from sea water, the exposure shall be classified as follows:
Note: Concerning moisture conditions, see also section 2 of this table.

XD1	Moderate humidity	Concrete surfaces exposed to airborne chlorides
XD2	Wet, rarely dry	Swimming pools Concrete exposed to industrial waters containing chlorides
XD3	Cyclic wet and dry	Parts of bridges exposed to spray containing chlorides Pavements Car park slabs

4 Corrosion induced by chlorides from sea water
Where concrete containing reinforcement or other embedded metal is subject to contact with chlorides from sea water or air carrying salt originating from sea water, the exposure shall be classified as follows:

XS1	Exposed to airborne salt but not in direct contact with sea water	Structures near to or on coast
XS2	Permanently submerged	Parts of marine structures
XS3	Tidal, splash and spray zones	Parts of marine structures

5 Freeze–thaw attack with or without de-icing salts

Where concrete is exposed to significant attack by freeze–thaw cycles while wet, the exposure shall be classified as follows:

XF1	Moderate water saturation, without de-icing agent	Vertical concrete surfaces exposed to rain and freezing
XF2	Moderate water saturation, with de-icing agent	Vertical concrete surfaces of road structures exposed to freezing and airborne de-icing agents
XF3	High water saturation, without de-icing agent	Horizontal concrete surfaces exposed to rain and freezing
XF4	High water saturation, with de-icing agent or sea water	Road and bridge decks exposed to de-icing agents Concrete surfaces exposed to direct spray containing de-icing agents and freezing Splash zone of marine structures exposed to freezing

6 Chemical attack

Where concrete is exposed to chemical attack from natural soils and ground water as given in Table 2 of EN 206-1, the exposure shall be classified as given below. The classification of sea water depends on the geographical location, therefore the classification valid in the place of use of the concrete applies.

Note: A special study may be needed to establish the relevant exposure class where there is:
– limits outside of Table 2 of EN 206-1;
– other aggressive chemicals;
– chemically polluted ground or water;
– high water velocity in combination with the chemicals in Table 11.2 of EN 206-1.

XA1	Slightly aggressive chemical environment according to Table 11.2 of EN 206-1
XA2	Moderately aggressive chemical environment according to Table 11.2 of EN 206-1 or exposure to sea water
XA3	Highly aggressive chemical environment according to Table 11.2 of EN 206-1

This system breaks the exposure environment down into the different major degradation factors that affect durability of the concrete and its reinforcement:

- carbonation-induced corrosion
- chloride-induced corrosion of reinforcement (treated separately for sea water and de-icing salts)
- freezing and thawing
- chemical attack

Each of these factors is then divided into sub-clauses by severity, dependent on moisture conditions and other influential factors.

The exposure environment is characterized for an individual element or part of the structure by an overall exposure class made up of all the relevant individual exposure classes and their level of severity, for example XC4 + XS3 + XA2 + XF4 for a pier on a marine jetty in sub-arctic conditions, or simply XC1 for an internal column in a heated building.

The exposure classification system in EN 206 has resulted from the great increase in knowledge about mechanisms of concrete deterioration and the increase in choice of methods to provide resistance. Specification of the exposure environment as, say, XC4 + XD3 + XF4, clearly indicates to the designer or concrete technologist the true nature of the environment in relation to its effect on the concrete and its reinforcement. This then enables the solution to be tailored through the choice of concrete, and possibly other protective measures, to these actions.

This exposure classification system is a great improvement over many older systems such as that in the British Standard code of practice for structural concrete (BS 8110-1, 1997). It classifies the environment by the effect it has on the concrete and its reinforcement. The classification system of BS 8110, as shown in Table 11.3, is more general and often combines several effects into one class. The exposure class 'very severe', for example, includes:

- exposure to chlorides and the consequent risk of reinforcement corrosion
- exposure to aggressive chemicals and the risk of loss of concrete section through deleterious reactions
- exposure to freezing and thawing with the risk of loss of concrete section through spalling or scaling.

Each of these types of degradation requires a different approach to providing adequate durability and the simple specification of 'very severe' exposure is not helpful to the concrete technologist in providing the best solution (chloride impermeability, chemical resistance, air entrainment or an appropriate combination of these, for example).

Table 11.3 Exposure classification system from the British Standard code of practice for structural concrete (BS 8110-1, 1997)

Environment	Exposure conditions
Mild	Concrete surfaces protected against weather or aggressive conditions
Moderate	Exposed concrete surfaces but sheltered from severe rain or freezing while wet Concrete surfaces continuously under non-aggressive water Concrete in contact with non-aggressive soil Concrete subject to condensation
Severe	Concrete surfaces exposed to severe rain, alternate wetting and drying or occasional freezing or severe condensation
Very severe	Concrete surfaces occasionally exposed to sea water spray or de-icing salts (directly or indirectly) Concrete surfaces exposed to corrosive fumes or severe freezing conditions while wet
Most severe	Concrete surfaces frequently exposed to sea water spray or de-icing salts (directly or indirectly) Concrete in sea water tidal zone down to 1 m below lowest low water
Abrasive	Concrete surfaces exposed to abrasive action, e.g. machinery, metal tyred vehicles or water carrying solids

The American building code (ACI 318-02) does not have an exposure classification system, as such, but identifies conditions which require special attention as shown in Table 11.4. The same general approach is taken in the Japanese architectural standard specification (JASS-5, 1997).

Table 11.4 Requirements for special exposure conditions from the American building code (ACI 318-02)

Exposure condition	Maximum free water–cementitious ratio, by weight, normal weight aggregate concrete	Minimum f'_c, normal weight and lightweight aggregate concrete	
		Psi	MPa
Concrete intended to have low permeability when exposed to water	0.50	4000	27.5
Concrete exposed to freezing and thawing in a moist condition or to de-icing chemicals	0.45	4500	31
For corrosion protection of reinforcement in concrete exposed to chlorides from de-icing chemicals, salt, salt water, brackish water, seawater, or spray from these sources	0.40	5000	34.5

11.5.6 Resistance to degradation

General

Standards and codes of practice for design of concrete structures, or for concrete itself, contain requirements or recommendations for the minimum quality of concrete and, sometimes, other protective measures for the defined exposure class. These standardized provisions eliminate the need for detailed knowledge of concrete technology for designers working in most common exposure conditions. Provided the exposure has been classified correctly, compliance with the requirements in the appropriate standard will provide a 'deemed-to-satisfy' solution with a generally acceptable degree of reliability. This is not necessarily the only solution and, in some circumstances, may provide insufficient reliability for the particular application. It should always be remembered that the recommendations or requirements in standards or codes of practice are usually the minimum believed necessary to achieve adequate performance. It is generally possible to work outside of the recommendations of the appropriate code or standard, depending on the legal status of these documents in the place of the works, but the designer may need to assume responsibility for the alternative approach. In some countries, such as France, failure to comply with appropriate codes may make it impossible to obtain essential buildings insurance.

Specification of durability by strength grade

In the days before autobatch records and quality schemes for readymixed concrete it was difficult for purchasers of concrete to satisfy themselves of the composition of the concrete supplied into their works in relation to their specification requirements for durability (i.e. minimum cement content, maximum free water–cement ratio). Indeed, this remains so in some parts of the world. The structural requirement for a minimum compressive strength is, however, relatively easy to check and, indeed, tends to be routinely done so. This provided a suitable means for specifying and checking durability requirements:

For a given cement type, durability in many exposure environments is generally inversely proportional to free water–cement ratio. Strength is, of course, also inversely proportional to free water–cement ratio so, for given materials, durability can in many exposure environments be related to compressive strength class. Provided a conservative relationship between strength and free water–cement ratio is known for the given set of materials, compliance with a specified strength grade can provide assurance that the required limits on maximum free water–cement ratio have not been exceeded. Prior to the use of water-reducing admixtures it was also possible also to confirm cement content by this method. Thus the reason behind specification of durability by strength is not a belief that durability is indeed controlled by strength but that essential requirements such as having a suitable maximum free water–cement ratio can be ensured this way.

Use of fly ash, ground granulated blastfurnace slag and other additions

Traditionally within British Standards the influence of fly ash (pulverized-fuel ash), and ground granulated blastfurnace slag (ggbs) as part of the cement has been treated on the basis of equivalence of concrete strength with Portland cement concrete. These Standards contained requirements for minimum cement content, maximum free water–cement ratio and minimum compressive strength grade originally developed for Portland cement concretes. As fly ash and ggbs cement concretes gained popularity these requirements were extended to these concretes but the values remained the same. Adequate durability relied on the assumption that, for example, where C40 concrete was required for resistance to chloride ingress, at the specified depth of cover to reinforcement, acceptable performance would be achieved using a 30 per cent fly ash cement C40 concrete just as it would be for a C40 Portland cement concrete. The exception to this was exposure to chemically aggressive environments such as sulfate-bearing groundwaters. In this case no requirement for minimum strength was given but the minimum cement content and maximum free water–cement ratio were individually tailored to three groups of cement, based on their composition and consequent durability performance.

This latter approach has now been adopted within the complementary British Standard to the European Standard for concrete (BS 8500, 2002). The regulatory framework for this standard dictated that requirements for concrete should continue to be given for minimum cement content, maximum free water–cement ratio and, optionally, minimum strength class but these have again been tailored to individual groups of cement depending on composition. This change of approach was the direct application of the vast amount of research information on the performance of concrete containing ground granulated blastfurnace slag and fly ash. This research had shown that although the carbonation performance of concrete containing different cement types was generally related to compressive strength class, other properties such as chloride ingress were dependent on other parameters, including free water–cement ratio. BS 8500 attempts to reflect these differences in its requirements.

In the future it may be possible to remove the need to distinguish between different cement types in these ways by the use of performance-based standards which will not rely on prescriptive requirement for composition.

Carbonation-induced corrosion of reinforcement

Resistance to carbonation-induced corrosion of reinforcement is generally provided in standards by the specification of a minimum thickness of concrete cover of specified quality. The quality of the concrete is generally specified in terms of prescriptive composition limits such as minimum cement content, maximum free water–cement ratio, minimum strength grade or a combination of two or more of these parameters.

It seems logical that if a higher than the permitted minimum quality of concrete is used then a lower depth of cover should be needed to provide the same level of protection and this is reflected in British Standards. This allows the designer to make a choice to 'trade-off' cover against increased concrete quality. This choice, curiously, is not available within most other European or other countries' standards.

The minimum concrete quality and/or the minimum cover to reinforcement are related to the degree of risk of deterioration identified by the exposure classification. Such measures should be adequate in terms of materials specification to resist carbonation-induced corrosion of reinforcement as research has repeatedly shown that most problems due to this type of deterioration can be attributed to provision of less than the specified depth of cover, or failure to use or compact the required concrete. This has not always been the case and current standards in humid countries, where this exposure condition can be problematic, have increased the required depth of cover and/or quality of concrete gradually over the years.

Chloride-induced corrosion of reinforcement

Similarly to carbonation-induced corrosion of concrete, most standards base the protection strategy against chloride-induced corrosion of reinforcement on the provision of sufficient depth of cover of concrete of adequate quality. Again quality is generally dictated by requirements for minimum cement content, maximum free water–cement ratio and minimum compressive strength. Nevertheless, in the past performance in severe chloride-bearing environments has on many occasions has been less than expected. Although this has often been the result of poor detailing, particularly with respect to water run-off, attention has also been focused on the adequacy of the requirements for cover and concrete quality and the possible need for other measures. Few standards mention the need, or give requirements, for other protective measures, such as surface treatments or corrosion-resistant reinforcement, to guard against chloride-induced corrosion.

Some *specifications* have left prescriptive requirements for limits on concrete composition behind and give their requirements in terms of a limit on chloride diffusivity, water permeability or other parameters believed to be related to chloride ingress. While admirable in their philosophy some such specifications have been written without a full understanding of the problem. This has resulted in cases where the specified limits have been almost impossible to achieve in practice. Values based on chloride ingress models should be used with caution unless the validity and limitations of the particular model are well understood.

Recent development of the complementary British Standard to the European Standard for concrete (BS 8500, 2002) involved a review of much of the available research to establish the state of the art with respect to chloride-induced corrosion. Although the regulatory framework for the standard meant that a prescriptive approach had to be used in preference to a performance-based specification, several improvements were made over many existing standards. The improved definition of exposure environment in the European standard (EN 206-1, 2000), as described in 4.5, allows the protective requirements to be tailored closely to the actual problem. Also the requirements were related to a specific lifetime. Indeed, as the standard only gives specific recommendations in terms of concrete quality and cover to reinforcement and not additional protective measures, it has been decided to limit the recommendations to a design life of 50 years. For longer design lives it is suggested that each design should be considered on a case-by-case basis and that other protective measures should be considered such as reducing contact between the chloride and the concrete, or the use of stainless steel reinforcement.

Where recommendations for concrete quality and depth of cover are given, they are specifically related to the degree of exposure and the type of cement. Cement type is divided into three groups depending on the amount of additions such as fly ash and ground granulated blastfurnace slag. Less onerous values of minimum cement content and maximum free water–cement ratio are required for cements containing suitable amounts of these materials because of their improved resistance to chloride ingress.

Limits of chloride content of concrete

Limits on the chloride content of fresh concrete containing reinforcement or other embedded metal are included for at least two reasons:

- To guard against the use of chloride-based admixtures which have been the cause of reinforcement corrosion in the past. At the time when these materials were used, research was reported to have shown that chloride, generally in the form of calcium chloride used as an accelerator, was safe in the intended dosages and would not lead to reinforcement corrosion. Experience proved this to be untrue although many problems could probably be attributed to excessive chloride dosages and poor distribution within the concrete leading to locally high concentrations.
- To ensure the time for chloride levels to reach threshold levels at the reinforcement is not significantly reduced by having an initially high level in the concrete from the mix constituents. Ideally the level of chloride would be zero but this is impossible in practice so the levels have been set at the lowest possible while still being practically achievable.

Until recently in the UK the initial limit for chloride was lower for concrete containing sulfate-resisting Portland cement than for other cement types but this was changed to make it the same as other cements following research that suggested the difference was unfounded. The limit for prestressed concrete is lower than that for normally reinforced concrete, not because prestressing strand is more prone to corrosion but because of the generally higher consequences of failure.

Freezing and thawing attack

Resistance to freezing and thawing attack is generally provided in standards by the requirement of a certain proportion of entrained air. Experience has shown that in less severe conditions adequate performance can be achieved by non air-entrained concrete of adequate strength and some standards such as BS8110 (BS 8110-1, 1997) and BS 8500 (BS 8500, 2002) in the UK allow this. Different values for the minimum permitted air content of concrete are often specified for various aggregate sizes to reflect the different proportions of cement paste in the concrete as seen in Table 11.5 taken from the American building code (ACI 318-02).

Research has shown that the resistance to freezing and thawing is dependent on the entrained air being in the form of small bubbles at close spacing, and some standards contain limits for these parameters. Some other standards require the air-entraining admixtures themselves to meet specified criteria for provision of suitable air void parameters; the performance in the concrete is then assumed to be adequate provided the admixture complies with these requirements. Standards generally contain limits on concrete

Table 11.5 American building code (ACI 318-02) requirements for entrained air content dependent on aggregate size and severity of exposure

Nominal maximum aggregate size (in)	Air content (%)	
	Severe exposure	Moderate exposure
3/8	7.5	6
1/2	7	5.5
3/4	6	5
1	6	4.5
$1\frac{1}{2}$	5.5	4.5
2	5	4
3	4.5	3.5

composition, such as maximum free water–cement ratio and minimum strength class, in addition to air entrainment.

Research has shown that not all cement types give equal performance against freezing and thawing attack and de-icer scaling, even when air-entrained, so standards may contain limits on cement composition, such as shown in Table 11.6 taken from the American building code (ACI 318-02).

Table 11.6 American building code (ACI 318-02) requirements for limits on cement composition for use in concrete exposed to de-icing chemicals

Cementitious materials	Maximum per cent of total cementitious materials by weight
Fly ash or other pozzolans conforming to ASTM C618	25
Slag conforming to ASTM C989	50
Silica fume conforming to ASTM C1240	10
Total of fly ash or other pozzolans, slag and silica fume	50
Total of fly ash or other pozzolans and silica fume	35

Sulfate attack

Research has shown that resistance to sulfate attack is influenced both by the quality of the concrete and the composition of the cement. Some cements are classified as sulfate-resisting but are only truly effective when used in a concrete of sufficiently low free water–cement ratio and adequate cement content for the particular degree of severity of the exposure. The new British Standard for concrete (BS 8500, 2002) acknowledges the good sulfate resistance of cement containing at least 25 per cent by mass of fly ash or at least 70 per cent by mass of ground granulated blastfurnace cement, as well as sulfate-resisting Portland cement, in all levels of sulfate exposure. The American building code (ACI 318-02), however, permits the use of fly ash and slag cements in moderate sulfate conditions but requires sulfate-resisting Portland cement (type V) in severe conditions, and a combination of type V cement and fly ash in very severe conditions.

Requirements related to sulfate attack in the UK have been updated several times in the past few decades as more research data have become available, particularly on the use of fly ash and slag cements. The most recent update has been made following the discovery of the thaumasite form of sulfate attack in a number of structures. The current standard (BS 8500, 2002) contains requirements based largely on the results of research into this form of attack, but which are quite complicated compared to those of many other countries' standards. These requirements contain limits on minimum cement content and maximum free water–cement ratio, dependent on the particular cement type used and on the carbonate content of the aggregate. In more severe exposure conditions the requirements relating to concrete composition are supplemented by the need for 'additional protective measures' such as drainage, a sacrificial layer of concrete, or protective coating. These measures reflect the performance of concrete in laboratory tests and from observation of affected structures. This research is continuing and it is expected that new results will lead to future developments in the recommendations.

Acid and other chemical attack

Research has generally shown that resistance to acid attack is related to the impenetrability of the concrete rather than the particular cement type used. Requirements for resistance to acid attack are thus often expressed simply in terms of limitations on concrete composition such as minimum cement content or maximum free water–cement ratio.

Resistance to other chemical attack is often only required in specialist applications such as chemical plants, or in contaminated land. The conditions may vary widely and most standards either do not address such issues or recommend the user to seek specialist advice.

Alkali–aggregate reaction

Research into the nature and avoidance of alkali–aggregate reactions in concrete has been very extensive over at least the last two decades. Standards have been updated as the state of knowledge has developed and in response to the need for simplicity in their requirements.

In the UK the latest requirements are given in the new British Standard for concrete (BS 8500-2, 2002) and vary according to the assessed reactivity of the aggregate (low, normal or high reactivity, or greywacke-type aggregate). These requirements are based directly on the results of research with UK materials and allow minimization of risk by measures including:

- Limits on cement content
- Limits on alkali content of cement
- The use of fly ash (pfa) or ground granulated blastfurnace slag

The standard also allows for use without limitation where there is a satisfactory history of use of the particular combination of coarse and fine aggregate, for at least ten years, in wet conditions with at least the same content of cement of the same alkali level or higher. Requirements relating to highly reactive aggregates and the use of other measures such as silica fume, metakaolin and lithium compounds are not contained in the standard but reference is made to BRE Digest 330 (BRE, 1999) and the Concrete Society Technical Report 30 (CSTR 30, 1999). No requirements are given in UK standards for avoidance of alkali–carbonate reaction as this is not perceived as a sufficiently significant problem there.

Recommendations in the USA are not contained in their building code (ACI 318-02) or specifications for structural concrete (ACI 301-99) but are considered in some depth in the ACI guide to durable concrete (ACI 201.2R-92). This document recommends that, in the absence of satisfactory performance requirements, reactivity of suspected aggregates should be assessed by standard laboratory tests including petrographic examination (ASTM

C295-98), mortar bar expansion (ASTM C227-97) and chemical tests (ASTM C289-01). Measures that are recommended for avoidance of reaction of potentially reactive aggregates include:

- Avoidance of their use in sea water or other environments where alkalis could enter from an external source
- Use of low-alkali cement
- Use of fly ash or slag

Recommendations are also given for avoidance of alkali–carbonate reaction which, probably because of the availability of less research information than for alkali–silica reaction, focuses solely on avoidance of rock identified as potentially reactive.

Abrasion

Abrasion can occur under a wide variety of conditions but recommendations in standards tend to be restricted to general statements or deal only with the more common occurrences such as floors. Recommendations in the American Concrete Institute guide to durable concrete (ACI 201.2-92) include:

- Selection of appropriate concrete strength
- A low water–cement ratio at the surface
- Proper grading of coarse and fine aggregate
- Use of the lowest slump compatible with proper placement and compaction
- High-strength toppings
- Special aggregates
- Proper finishing techniques
- Vacuum de-watering
- Special dry-shakes and toppings
- Proper curing

The British Standard for concrete floors (BS 8204, 1999) contains recommendations for limits on minimum cement content, maximum free water–cement ratio and minimum strength grade as well as the use of special toppings depending on the required class of abrasion resistance.

11.5.7 Specification, production and conformity

Standards such as the British and European standards for concrete (BS 5328, 1997; EN 206-1, 2000) contain requirements for specification, production and conformity testing of concrete. Such standards can contribute to durability in various ways, such as by ensuring that:

- all the required information is included in a specification for concrete in a standard and unambiguous way
- concrete batching equipment is regularly checked and calibrated to ensure that the concrete contains the intended proportions of constituent materials
- mixing and transportation is done in a way that ensures the concrete properties are not adversely affected
- concrete production is tested at regular intervals to ensure compliance with the specification.

11.5.8 Fresh concrete test methods

It is important that concrete has properties in the fresh state suited to the method of distribution and placement. In order that the requirements of the constructor for workability can be communicated to the supplier, and that they can be checked prior to placement, it is essential to have standard test methods such as slump or flow.

Although many tens of workability test methods have been proposed over the years (Tattersall, 1976) it is interesting that the number that have been standardized are few (e.g. slump, flow/spread, compaction factor, compaction (Walz), VeBe, Kelly Ball) and that the number in common use is probably even less. Standardization of many techniques which purport to measure essentially the same property is undesirable and it is preferable that there should be as few as possible tests. That is not to say that new tests should not be developed but old ones should then be withdrawn. Tests such as the slump test may not be particularly scientific but they have the great advantage of almost universal adoption.

11.5.9 Hardened concrete test methods

Standards exist in most countries to cover routine tests such as compressive strength testing of standard test specimens, usually cubes or cylinders. Although durability is not related directly to compressive strength, there is usually a relationship between the concrete quality parameters that affect durability and strength for given materials. Strength testing has therefore been traditionally used as a check on durability. Indeed, a low strength may often be the result of an increased free water–cement ratio which, in turn, may adversely affect potential durability. Strength in a structure, measured on extracted cores, can have the added advantage of assessing the influence of execution. It is essential therefore that such test methods are standardized to ensure results of these tests are comparable and meaningful.

Assessment of current condition or residual durability at some stage in the life of a structure may rely on tests such as carbonation depth, chloride ion content, covermeter surveys, and any of a large array of non-destructive tests. The process of standardization of these techniques may lag many years behind developments in testing technology and standards may thus be unavailable for the latest test methods. Nevertheless, wherever possible, testing should be performed in accordance with the relevant standard or an appropriate authoritative specification to ensure a meaningful and useful outcome.

11.5.10 Execution, or workmanship

As stated elsewhere in this chapter and throughout this book, durability depends not just upon design, detailing and use of appropriate materials but also on satisfactory standards of execution, or workmanship as it is also known. Particular examples where execution has a large influence include:

- Achievement of specified cover to reinforcement, in particular to guard against chloride- and carbonation-induced corrosion of reinforcement

- Curing, particularly of unformed surfaces, to ensure adequate hydration in the near-surface and surface layers
- Finishing of horizontal and other unformed surfaces to ensure, for example, a dust-free surface with the required degree of abrasion resistance

Standards for execution, or clauses within other standards, have been developed in many countries in an attempt to ensure this part of the durability process receives the attention it deserves. A European Standard for execution of concrete structures was under development at the time of writing and was available as a prestandard (ENV 13670-1, 2000). This is a very large document (60 pages) containing requirements across all aspects of execution including:

- Falsework and formwork
- Reinforcement
- Prestressing
- Concreting (specification, handling fresh concrete, placing, compacting, curing, etc.)
- Precast elements
- Geometrical tolerances
- Inspection

Typical clauses, for example, for placing and compacting include:

> The concrete shall be placed and compacted in order to ensure that all reinforcement and cast-in items are properly embedded within the tolerances on the cover in compacted concrete and that the concrete achieves its intended strength.
>
> The rate of placing and compaction shall be high enough to avoid cold joints and low enough to prevent excessive settlements or overloading of the formwork and falsework.
>
> Segregation shall be minimized during placing and compaction.
>
> The concrete shall be protected against adverse solar radiation, strong wind, freezing, water, rain and snow during placing and compaction.

11.5.11 Concrete products

Many concrete products, such as pipes, cladding and piles, are covered by their own specific standards. These may contain specific requirements for composition, or particular minimum performance levels measured by specified tests, related to the particular conditions of use of that type of product. This allows the specifier or user to select products that are certified in accordance with such standards safe in the knowledge that they are suitable for their intended use.

11.5.12 Some limitations of standards

Standards can only contain requirements within the scope of coverage of that particular document even though satisfactory performance in service can usually only be obtained by also paying attention to factors outside that scope. An example of this is resistance to chloride-induced corrosion of reinforcement. Standards may give stringent requirements for limits on concrete composition but experience from service has shown that performance may be critically dependent on detailing of run-off of water laden with de-icing salts.

Durability can rarely be provided by the quality of concrete alone, which may be in one standard, or cover to reinforcement, which may be in another standard, but may also depend on detailing and standards of execution, which indeed may not be covered in any standard at all. Another example is resistance to abrasion which can depend on finishing techniques and curing at least as much as on the composition of the concrete.

Standards relating to concrete are tending to become larger and more complicated and it is becoming increasingly difficult for practising engineers to be fully aware of all new developments. There is a need for improved knowledge management techniques to assist in keeping up to date.

Many test method standards suffer limitations through the essential need to standardize procedures into a relatively simple and rapid test method. Generally, durability-related test methods only address a single deterioration mechanism, or related parameter, whereas concrete in service is typically subject to several degradation factors simultaneously. A road pavement in a cold, cool or temperate climate, for example, needs to resist both abrasion and freeze–thaw. These properties tend to be tested separately but, in practice, the effects of abrasion are worsened by freeze/thaw (Leshchinsky and Lesinskij, 2002). The actual performance in service may thus be worse than that predicted by the individual test methods.

These examples are intended not to be critical of individual standards or of standards writers but to highlight that the standardization process is inherently limited. Users of standards should be aware of this and be prepared to make allowance where necessary.

11.6 Specifications

11.6.1 Introduction

Specifications are an extremely important part of any construction project. They are the means by which specifiers pass their requirements on to the contractor or supplier. They may take the form of standard specifications (such as the National Structural Concrete Specification in the UK (NSCS, 2000) or they may be project- or company-specific. This section is concerned with the latter two types.

Specifications are usually legally binding as part of the construction contract. They may include reference to standards, codes of practice or general specifications which are not, in their own right, obligatory but their inclusion in the specification then endows them with an obligatory status within that contract.

Many specifications are very large documents and thus may not receive the attention they deserve until it is too late (e.g. when a problem has already occurred). The concrete specification for a recent major European marine rail and road link, for example, was 110 pages long. It is in all parties' interests to keep the specification as brief as is practical.

If a contractor or supplier is unable to comply with a specification they should contact the specifier immediately and seek to obtain a written variation to the requirements.

11.6.2 Typical specification clauses

General

> Unless modified by this specification the procedures to be used in producing, transporting, sampling and testing of the concrete shall comply with BS5328 : Parts 3 and 4. References are also made to individual clauses of BS 5328 : Part 1, and the concrete mixes are specified in accordance with the methods given in BS 5328 : Part 2. In cases of conflict, this specification takes precedence over BS 5328.

This shows that an effort has been made to minimize the size of the specification and keep it simple by reference to existing national standards. It also points out that there are some modifications to those standards and makes clear the relationship between the specification and the standard by stating which takes preference.

Sampling for strength

> Samples shall be taken at the point of discharge from the mixer or delivery vehicle, or at the point of placing the concrete, as directed.
> The rate of sampling shall be in accordance with clause 9.2 and table 15 of BS 5328: Part 1: 1997.
> In addition, for each sample, one cube shall be made for testing at 7 days.

The extra cube is not for formal checking of conformity so this requirement would not need to be passed on to the concrete supplier. Its purpose is generally to provide advance warning of potentially low 28-day strength values.

Temperature monitoring and control

> When concrete is to be placed in large volume pours or when rich mixes are to be placed at high ambient temperatures, the temperature of the concrete shall be monitored through the section.
> The contractor shall ensure that the temperature of the concrete does not exceed 65°C.
> The temperature gradient across the section shall be controlled by the provision of thermal insulation to prevent a differential greater than 20°C.

This clause is performance related, rather than prescriptive, insomuch as it gives limits for maximum temperature and temperature difference across a section. These are necessary to avoid lower than expected ultimate strength, the risk of delayed ettringite formation, and early thermal contraction cracking due to internal restraint. It leaves the freedom of how to achieve these requirements up to the constructor.

The third specification requirement could, however, be improved by allowing a different temperature difference to be used if the concrete is made using coarse aggregate of low coefficient of thermal conductivity.

11.6.3 Writing specifications

- Keep specifications as brief as possible
- Only specify what *needs* to be specified
- Where possible, refer to standard specifications, and national or international standards
- Make specification clauses clear, unambiguous, purposeful and achievable
- Ensure cited references are up to date
- Make clear where the specification deviates from or modifies national or international standards, or normal practice
- Be reasonable – unreasonable clauses may not be enforceable in law
- When working in a different country, where possible, avoid references to standards or other documents unfamiliar in that location

11.6.4 Using specifications

- Read them! – particularly if they are not a standard specification
- Comply with them
- Identify deviations from or modifications to standards or normal practice
- Query any ambiguities
- Identify potential problems and attempt to resolve them as soon as possible – alternatives may be acceptable but need to be approved by the specifier
- Ensure they are seen by the people that need to see them (e.g. specific mix requirements need to be seen by the concrete producer) and ensure that all relevant information is passed on
- Keep copies of standards and standard specifications up to date – they should be controlled copies within a QA system

References

ACI 201.2R (1992, reapproved 1997) Guide to durable concrete, American Concrete Institute.
ACI 301 (1999) Specifications for structural concrete, American Concrete Institute.
ACI 303.1 (1997) Standard specification for cast-in-place architectural concrete, American Concrete Institute.
ACI 318 (2002) Building Code requirements for structural concrete, American Concrete Institute.
ACI 349.1 (1991, reapproved 2000) Reinforced concrete design for thermal effects on nuclear power plant structures, American Concrete Institute.
Architectural Cladding Association (1998) Code of practice for the safe erection of precast concrete cladding, British Precast Concrete Federation, Leicester, UK.
ASTM C 227 (1997) Standard test method for potential alkali reactivity of cement aggregate combinations (mortar bar method), American Society for Testing and Materials.
ASTM C 289 (2001) Standard test method for potential alkali–silica reactivity of aggregates (chemical method), American Society for Testing and Materials.
ASTM C 295 (1998) Standard practice for petrographic examination of aggregates for concrete, American National Standard, American Society for Testing and Materials.
ASTM C1157 (2000) Standard performance specification for hydraulic cement, American Society for Testing and Materials.

BRE (1999) Digest 330, Alkali–silica reaction in Concrete – Parts 1–4, Building Research Establishment, Watford, UK.
BS 5328-1 (1997) Concrete – Part 1: Guide to specifying concrete, British Standards Institution.
BS 5328-2 (1997) Concrete – Part 2: Methods for specifying concrete mixes, British Standards Institution.
BS 5328-3 (1990) Concrete – Part 3: Specification for the procedures to be used in producing and transporting concrete, British Standards Institution.
BS 5328-4 (1990) Concrete – Part 4: Specification for the procedures to be used in sampling, testing and assessing compliance of concrete, British Standards Institution.
BS 5400-4 (1990) Steel, concrete and composite bridges – Part 4: Code of practice for design of concrete bridges, British Standards Institution.
BS 5502 (1990) Buildings and structures for agriculture, British Standards Institution.
BS 6349-1 (2000) Maritime structures – Part 1: Code of practice for general criteria, British Standards Institution.
BS 6744 (2001) Stainless steel bars for the reinforcement of and use in concrete – requirements and test methods, British Standards Institution.
BS 7973-1 (2001) Spacers and chairs for steel reinforcement and their specification – Part 1: Product performance requirements, British Standards Institution.
BS 7973-2 (2001) Spacers and chairs for steel reinforcement and their specification – Part 2: Fixing and application of spacers and chairs and tying of reinforcement, British Standards Institution.
BS 8007 (1987) Code of practice for design of concrete structures for retaining aqueous liquids, British Standards Institution.
BS 8103 (1995) Structural design of low-rise buildings, British Standards Institution.
BS 8204 (1999) Screeds, bases and in-situ floorings, British Standards Institution.
BS 8110-1 (1997) Structural use of concrete – Part 1: Code of practice for design and construction, British Standards Institution.
BS 8500-1 (2002) Concrete – Complementary British Standard to BS EN 206-1 – Part 1: Method of specifying and guidance for the specifier, British Standards Institution.
BS 8500-2 (2002) Concrete – Complementary British Standard to BS EN 206-1 – Part 2: Specification for constituent materials and concrete, British Standards Institution.
BS EN 206-1 (2000) Concrete – Part 1: Specifications, performance production and conformity, British Standards Institution.
Clark, L.A., Shammas-Toma, M.G.K., Seymour, D.E., Pallett, P.F. and Marsh, B.K. (1997) How can we get the cover we need? *The Structural Engineer*, **76**, No. 17.
Concrete Society Technical Report No. 30 (1999) *Alkali–silica reaction – minimising the risk of damage to concrete*, 3rd edition, The Concrete Society, Slough.
EN 197-1 (2000) Cement – Part 1: Composition, specifications and conformity criteria for common cements, European Committee for Standardization.
EN 206-1 (2000) Concrete – Part 1: Specification, performance, production and conformity, European Committee for Standardization.
EN 934-2 (2001) Admixtures for concrete, mortar and grout – Part 2: Concrete admixtures – definitions, requirements, conformity, marking and labelling, European Committee for Standardization.
ENV1991-1 (1996) Eurocode 1: Basis of design and actions on concrete structures. Part 1: Basis of design, European Prestandard, European Committee for Standardization.
ENV 1992-1-1 (1992) Eurocode 2: Design of concrete structures – Part 1: General rules for buildings, European Prestandard, European Committee for Standardization.
ENV 13670-1 (2000) Execution of concrete structures Part 1: Common, European Prestandard, European Committee for Standardization.
Hobbs, D.W. (ed.) (1998) *Minimum Requirements for Durable Concrete*, British Cement Association.
Leshchinsky, A. and Lesinskij, M. (2002) Concrete durability, selected issues. Part 1. L'Industrie Italiana del cemento (in English), May, 444–454.

ISO 14654 (1999) Epoxy coated steel for the reinforcement of concrete, International Standards Organization.
JASS-5 (1997) Japanese Architectural Standard Specification, Chapter 5 – Reinforced concrete works, Architectural Institute of Japan.
NSCS (2000) *National structural concrete specification for building construction*, 2nd edition, British Cement Association.
Tattersall, G.H. (1976) *The workability of Concrete, A viewpoint publication*, Cement and Concrete Association, Slough.

Further reading

ACI 318R (2002) Commentary on Building Code requirements for structural concrete, American Concrete Institute.
CS109 (1996) Concrete Society Discussion Document, Developments in durability design & performance-based specification of concrete.

Index

28-day strength *see* Early-age and accelerated strength testing
Abrasion problems, 11/24
Accelerated strength testing of concrete, 3/1–16
  about accelerated strength testing, 3/1
  American Society for Testing and Materials (ASTM) Procedures (C684), 3/5–7
    Procedure A – moderate heating, 3/5
    Procedure B – thermal acceleration, 3/5–6
    Procedure C – autogenous curing, 3/6–7
    Procedure D – elevated temperature and pressure, 3/7
  Australian standards, 3/7
  British Standards (BS 1881/BS EN 12390-1):
    35°C method, 3/3
    55°C method, 3/4
    85°C method, 3/4
    procedures, 3/3
    test report – mandatory/optional information, 3/5
  Canadian standards, 3/7
  Denmark standards, 3/7
  Japanese research, 3/7–8
  principles, 3/2–3
  Russian standards, 3/7
  Thailand research, 3/7
  see also Early-age and accelerated strength testing; Strength-testing of concrete
Acceptance/compliance testing, 9/17–18
ACI (American Concrete Institute) Manual of Concrete Practice, 11/5
Acid attack on reinforced concrete, 6/8
Acoustic emission testing:
  applications and limitations, 6/41–2
  theory, 6/41
Additions, durability requirements, 11/12

Admixtures:
  durability requirements, 11/13
  with reinforced concrete, 6/3
Aggregates:
  drying shrinkage, 6/6
  durability requirements, 11/13
  grading for fresh concrete tests, 1/21–2
    buoyancy method, 1/21
    pressure filter (Sanberg) method, 1/2–21
    RAM method, 1/21–2
  grading for hardened concrete, 4/12
  for reinforced concrete, 6/3
Agrément Board schemes, 8/17
Alkali-aggregate reaction, degredation from, 11/23–4
Alkali-silica reaction (ASR) in reinforced concrete, 6/5, 6/14, 7/13
Alkalis content, 4/11
American building code, 11/4
  see also Accelerated strength testing of concrete; Standards
American Society for Testing and Materials (ASTM) *see* Accelerated strength testing of concrete
Analysis of concrete and mortar *see* Buoyancy test for cement content; Fresh concrete analysis; Fresh concrete sampling; Hardened concrete and mortar analysis; Pressure filter (Sandberg) cement content analysis method; RAM (rapid analysis machine) cement content (constant volume) test; Statistical analysis
Anti-carbonation paint with reinforced concrete, 7/5
ASR (Alkali-silica reactivity) in reinforced concrete, 6/5, 6/14, 7/13

Index

British standards *see* Standards
BRMCA (British Ready Mixed Concrete Association), 9/2
Buoyancy method for aggregate grading, 1/21
Buoyancy test for cement content, 1/6–8
    analysis procedure, 1/7
    calibration for relative densities, 1/6–7
    cement mass, 1/8
    coarse aggregate mass, 1/7–8
    fine aggregate mass, 1/7–8
    fines correction factor, 1/7
    relative density calibration, 1/6–7

Capillary porosity, 4/12
Carbon dioxide content, 4/13
Carbonation in reinforced concrete, 7/9, 11/15, 11/19–20
    depth of, 6/29–31
    repairing, 7/3–5
CARES scheme for reinforced concrete, 8/18, 8/19
Cathodic protection of reinforced concrete:
    basic chemistry, 7/6, 7/10
    conductive coating systems, 7/8
    corrosion process, 7/6–7
    impressed current cathodic protection, 7/8
    reactivity, 7/7
    Sacrificial flame- or arc-sprayed zinc protection system, 7/8
    sacrificial protection, 7/7
Cavity detection, with UPV measurements, 6/24
CE marks, 8/17–18
Cement content *see* Fresh concrete analysis; Hardened concrete and mortar analysis
Cement effects on concrete durability, 11/12
Cement with reinforced concrete:
    cement content effects, 6/29
    cement content test methods, 6/29
    high alumina cement, 6/2
    ordinary Portland cement, 6/2
    sulfate-resisting cement, 6/2
CEN (European Committee for Standardization), 11/6–8
Central Limit Theorem (CLT), 10/13–14, 10/15
Chemical attack classification, 11/16
Chloride attack on reinforced concrete, 7/9–10, 11/20–1
    and carbonation, 7/9
    effects, 6/27–8
    exposure classification, 11/15
    limit considerations, 11/21
    repair, 7/5
    test methods, 6/27
Chloride removal (CR) (chloride extraction and desalination):
    about CR, 7/9
    advantages/disadvantages, 7/15
    alkali-silica reactivity (ASR) acceleration, 7/13
    anode types, 7/11
    bond strength reduction problem, 7/13
    carbonation, 7/9
    case histories:
        Burlington, 7/14
        Tees, 7/14
    chloride attack process, 7/9–10
    electrolytes, 7/11
    end point determination, 7/12–13
    operating conditions, 7/12
Codes of practice:
    about codes of practice, 11/2
    ACI Manual of Concrete Practice, 11/5
    role and status, 11/3–5
    selection of, 11/4–5
Comparison of means, 10/18–20, 10/24
Comparison of variances, 10/20–1, 10/25
Compliance/acceptance testing, 9/17–18
Compression testing *see* Core sampling and testing; Strength-testing of concrete
Concrete products, durability, 11/26
Concrete Society Working Party (1971), 2/3–4
Conductive coating systems of protection, 7/8
Confidence levels, 10/16–17
Confidence lines, 10/33
Constant volume test *see* RAM (rapid analysis machine) cement content (constant volume) test
Control charts:
    about control charts, 9/2–3
    and confidence intervals, 10/17–18
    see also Cusum charts/technique; Shewhart charts
Core sampling and testing:
    about core taking and testing, 5/2–3
    and compression testing, 5/7
    conditioning of cores, 5/6–7
    core size considerations, 5/4
    curing, 5/17
    document guidance aspects, 5/4–5
    drilling considerations, 5/5
    in-situ cube strength, 5/3, 5/4, 5/8–11, 5/12–13, 5/16
    location of sampling points, 5/5, 5/6
    orientation factor, 5/17–18
    planning and preliminary work, 5/3–4
    Point Load Strength Test, 5/8
    potential strength, 5/3, 5/4, 5/8–10, 5/13, 5/14
    preparation of cores, 5/6–7
    reinforcement congestion problems, 5/4
    result interpretation, 5/10–12
    standards and guidance, 5/1–2
    visual examination and measurement, 5/5–6, 5/8
    voidage, 5/7, 5/13, 5/18, 5/19
    worked examples:
        concrete slab, 5/12–13
        office block ground floor, 5/14
    see also CSTR No. 11 update
Correlation/regression, 10/26–8
    correlation coefficient, 10/29–31
Corrosion *see* Reinforced concrete
Corrosion inhibitors (reinforced concrete), 7/18–19
Covermeter survey, 6/16–17
CR *see* Chloride removal (CR) (chloride extraction and desalination)
Crack estimation, with UPV measurements, 6/24–5
Cracking of reinforced concrete, 6/11–15

Index

CSTR No. 11 update:
   about CSTR No. 11 update, 5/15
   curing considerations, 5/17
   interpretation options, 5/19–21
   new data acquisition, 5/15–16
   new data results, 5/16–17
   orientation factor, 5/17–18
   voidage, 5/18, 5/19
Curing, and core sampling and testing, 5/17
Customer's and producer's risk, 9/19–20
Cusum charts/technique:
   28-day predictions monitoring, 9/12–13
   about Cusum charts, 9/8
   commercial implications, 9/16–17
   correlation tables, 3/14
   Cusum M, R and C plots, 9/11–15
   mean strength of concrete calculation and control, 9/9–11
   practical application, 9/15–16
   properties of Cusum system, 9/13–15
   QSRMC scheme, 9/8, 9/9
   standard deviation control, 9/11–12

Data representation, 10/3–5
DEF (delayed ettringite formation) in reinforced concrete, 6/12–15
Defect detection, with UPV measurement, 6/24
Degradation resistance:
   abrasion, 11/24
   alkali-aggregate reaction, 11/23–4
   carbonation-induced corrosion of reinforcement, 11/19–20
   chloride-induced corrosion, 11/20–1
   and durability by strength grade, 11/18–19
   fly ash effects, 11/19
   freezing and thawing attack, 11/21–2
   ground granulated blastfurnace slag effects, 11/19
   sulphate attack, 11/22–3
   see also Durability in standards and specifications
Desalination *see* Chloride removal (CR) (chloride extraction and desalination)
Drilling core samples, 5/5
Drying shrinkage, aggregates, 6/6
Durability in standards and specifications:
   about materials for durability, 11/11
   'additions' requirements (EN 206-1), 11/12
   admixture requirements, 11/13
   aggregate requirements, 11/13
   cement requirements (EN 197-1 and ASTM C1157-00), 11/12
   concrete products, 11/26
   constituent materials test methods, 11/14
   designing for, 11/10–11
   ENV 1991-1, 11/10–11
   and execution, or workmanship, 11/25–6
   exposure environment classification, 11/14–18
   strength grade correlation, 11/18–19
   water requirements (AC 318-99), 11/12
   see also Degradation resistance

Early-age and accelerated strength testing, 3/8–15
   about early-age testing, 3/8–10
   BS 1881, 3/11
   compared with high-temperature accelerated testing, 3/15–16
   conformity, 3/13–15
   control by prediction of 28-day strength, 3/10–13
   Cusum correlation tables, 3/14
   fixed set boiling water method, 3/10–11
   ggbs effects, 3/13
   King's procedure, 3/10
   Patch's method, 3/10
   see also Accelerated strength testing
Electrical resistivity reinforcement corrosion testing:
   equipment, 6/46
   interpretation, 6/46–7
   limitations, 6/47
Epoxy-coated reinforcement, 11/14
ERSG (electrical resistance strain gauge), 2/7
Ettringite formation, delayed in reinforced concrete, 6/12–15
Eurocodes, 11/7
European Committee for Standardization (CEN), 11/6–8
Expanded titanium mesh anode protection system, 7/8
Exposure environment classification:
   American Building Code (ACI 318-02), 11/17–18
   European standard EN 206, 11/14–17

F-distribution critical values table, 10/38–9
Fire damage:
   reinforced concrete, 6/9–10
   strength estimation with UPV measurements, 6/25–6
Fly ash, effect on degradation, 11/19
Footemeter strain-gauged column, 2/6–7
Freeze-thaw attack:
   classification, 11/16
   degradation effects, 6/5–6, 11/21–2
Fresh concrete analysis:
   about fresh concrete analysis, 1/3
   advantages/disadvantages, 1/3
   BS 1881, 1/4
   cement content, 1/4–14
      applicability of test methods, 1/5
      calibration samples, 1/4
      test samples, 1/4–5
   test method standardization, 11/25
   see also Buoyancy test for cement content; Fresh concrete sampling; ggbs (ground granulated blastfurnace slag) content tests; pfa (pulverised-fuel ash) content tests; Pressure filter (Sandberg) cement content analysis method; RAM (rapid analysis machine) cement content (constant volume) test; Water content of fresh concrete
Fresh concrete sampling:
   calibration samples, 1/4

Index

Fresh concrete sampling (*Continued*)
  from bulk concrete quantities, 1/5
  from mixing or agitating trucks, 1/5
  test samples, 1/4–5
Frost damage *see* Freeze-thaw attack

GECOR 6 device for reinforcement corrosion measurement, 6/49
ggbs (ground granulated blastfurnace slag) content tests, 1/17–19, 4/12–13
  calibration, 1/18–19
  chemical test apparatus and procedure, 1/17–18
  and early/accelerated strength testing, 3/13
  test validity, 1/19
ggbs (ground granulated blastfurnace slag) effect on degradation, 11/19

Half cell potential reinforcement corrosion testing:
  about half cell potential testing, 6/43–4
  concrete cover depth problems, 6/44
  concrete resistivity problems, 6/44
  measurement procedures, 6/45
  polarization effects, 6/44
  results and interpretation, 6/45
Hardened concrete and mortar analysis:
  about analysis, 4/2–3
  admixtures content, 4/13
  aggregate grading, 4/12
  alkalis content, 4/11
  calcium oxide content, 4/6–8
  capillary porosity, 4/12
  carbon dioxide content, 4/13
  cement content, 4/6–9
  cement content accuracy and precision of measurements, 4/13–14
  chloride content, 4/11
  ggbs content, 4/12–13
  history, 4/1–2
  microsilica and metakaolin considerations, 4/13
  mortar mix proportions, 4/9–11
    accuracy of measurements, 4/14
  pfa content, 4/13
  pfa problems, 4/9
  soluble silica content, 4/8
  sulfate content, 4/11
  test method standardization, 11/25
  water/cement ratio, 4/11–12
Hardened concrete and mortar sampling:
  core cutting, 4/4
  dust drilled samples, 4/4
  general rules, 4/3–4
  lump samples, 4/4
  mortar samples, 4/4–5
  number of samples, 4/5
  sample preparation, 4/6
  sampling strategy, 4/3–4
Health and safety, reinforced concrete repair, 7/5
High alumina cement, with reinforced concrete, 6/2
High alumina cement (HAC) structure problems:
  causes/effects, 6/31
  test methods, 6/31

Histograms, 10/4
Hypothesis testing, 10/21–3

In-situ and potential strength, 5/3, 5/4, 5/8–10, 5/12–14, 5/16
Incipient anode effect, reinforced concrete, 7/5–6
Infra-red thermography, 6/42–3
ISO (International Standards Organization):
  about ISO, 11/6
  ISO 9001: 2000:
    about ISO 9001, 8/5–10
    audit and review of management and procedures, 8/27–9
    consultation and communication, 8/12–13
    continuous improvement requirement, 8/6, 8/26–7
    customer focus, 8/11
    integrated management approach, 8/8
    laboratory management, 8/29–30
    management review, 8/22–3
    non-conformity, 8/25–6
    people involvement, 8/11–12
    procedure and method statements, 8/13–14
    process models, 8/5
    purchasing and supplier appraisal, 8/23–4
    reconciling ideas and text, 8/10–13
    risk management process, 8/13
    scope, application and definitions, 8/8
    system and product review, 8/22–3
    top management role, 8/6–7, 8/9
  ISO 17025 framework, 8/29–30

King's procedure, 3/10

Laboratory management, 8/29–30
Least-squares method, regression, 10/28–9
Linear polarization measurement of reinforcement corrosion rate:
  equipment and use, 6/48–9
  GECOR 6 device, 6/49
  interpretation of results, 6/50–1

Mortar analysis *see* Hardened concrete and mortar analysis

Normal distribution table, 10/35–6

Operating-characteristic (O–C) curves, 9/18–19
Ordinary Portland cement, with reinforced concrete, 6/2
Orientation factor, and core sampling, 5/17–18

Patch repairs of carbonation-induced corrosion, 7/3–5
Patch repairs of chloride-induced corrosion, 7/5
Patch's method, 3/10

Petrographic examination of reinforced concrete:
  broken surfaces, 6/34
  composition, 6/34
  polished surfaces, 6/33
  preliminary examination, 6/33
  thin sections, 6/33–4
  water/cement ratio, 6/34
pfa (pulverized-fuel ash) content tests, 1/14–17
  about pfa content tests, 1/14–15
  calibration, 1/15
  particle density:
    dry weight of material, 1/16
    wet weight of material, 1/15–16
  particle density calculation, 1/16–17
  test procedure, 1/17
Phenolphthalein spray method of steel reinforcement, 7/3–4
Plastic cracking, reinforced concrete, 6/11–15
Point Load Strength Test, 5/8
Population measures (statistical), 10/6
Portland cement, with reinforced concrete, 6/2
Potential and in-situ strength, 5/3, 5/4, 5/8–10, 5/12–14, 5/16
Pressure filter (Sandberg) cement content analysis method, 1/12–14
  and aggregate grading, 1/22
  analysis procedure, 1/13–14
  calibration, 1/12–13
  cement correction factor determination, 1/13
  coarse/fine aggregate mass, 1/13–14
  fines content determination, 1/12–13
  mass of each constituent per cubic metre of concrete, 1/14
  mass per cubic metre of fresh concrete, 1/14
  pfa content tests, 1/15, 1/16
  water absorption determination, 1/12–13
Probability/probability functions/value calculations, 10/6–10
Producer's and customer's risk, 9/19–20
Pulse velocity testing *see* UPV (ultrasonic pulse velocity) measurement from PUNDIT
PUNDIT *see* UPV (ultrasonic pulse velocity) measurement from PUNDIT

QSRMC (Quality Scheme for Ready Mixed Concrete), 8/19, 8/30, 9/8
Quality:
  about quality, 8/3–4, 9/1–2
  about sector schemes, 8/18–19
  about third-party registration and sector schemes, 8/16–17
  agrément board schemes, 8/17
  analysis of risk, 8/21
  CARES scheme for reinforced concrete, 8/19
  CE marks, 8/17–18
  compliance/acceptance testing, 9/17–18
  control charts, 9/2–3
  definitions, 8/3–4
  experimental design, 9/20–3
  operating-characteristic (O-C) curves, 9/18–19
  producer's and customer's risk, 9/19–20
  QSRMC scheme for readymixed concrete, 8/19
  self-certification and quality control, 8/20–2
  third-party accreditation and audit, 8/21
  see also Cusum charts/technique; ISO (International Standards Organization); Shewhart charts; Standards

Radar profiling:
  about radar profiling, 6/39
  applications, 6/40–1
  equipment, 6/40
  frequency modulation system, 6/39
  impulse radar system, 6/39
  limitations, 6/40–1
  synthetic pulse radar system, 6/39
RAM (rapid analysis machine) cement content (constant volume) test, 1/8–11
  about RAM method, 1/8
  and aggregate grading, 1/21–2
  analysis procedure, 1/11
  machine calibration, 1/9–11
  machine operation, 1/8–9
Random variations, 10/2–3
Realkalization (ReA):
  about realkalization, 7/15–16
  advantages, 7/17–18
  anode types, 7/16–17
  case histories, 7/17
  chemical process, 7/10
  electrolytes, 7/17
  end point determination, 7/17
  operating conditions, 7/17
Rebound hammer *see* Schmidt Rebound Hammer
Regression models:
  beam deflection example, 10/31
  confidence lines, 10/33
  correlation, 10/26–8
  correlation coefficient, 10/29–31
  least-squares method, 10/28–9
  multivariate problems, 10/33
  regression curve fit, 10/33
  residuals analysis, 10/31–2
Reinforced concrete structures, formation, problems and faults, 6/1–12
  about reinforced concrete structures, 6/1–2
  about structural failure, 6/4
  acid attack, 6/8
  admixtures for, 6/3
  aggregates for, 6/3
  alkali-aggregate reaction, 11/23–4
  alkali-silica reaction, 6/5, 6/14
  carbonation, 7/9, 11/15, 11/19–20
  cement types, 6/2
  chemical attack classification, 11/16
  chloride attack, 7/9–10, 11/15, 11/20–1
  cracking, 6/11–15
  DEF (delayed ettringite formation), 6/12–15
  exposure classification system, 11/15–18
  fire damage, 6/9–10
    dehydration of the cement hydrates, 6/9
    phases alteration in aggregate and paste, 6/9, 6/10
    surface cracking and microcracking, 6/9, 6/10

Index

Reinforced concrete structures, (*Continued*)
    fly ash usage, 11/19
    freeze-thaw attack:
        classification, 11/16
        degredation, 11/21–2
    frost damage, 6/5–6
    plastic settlement cracks, 6/11, 6/12, 6/14
    plastic shrinkage cracks, 6/11, 6/14, 6/15
    poor construction problems, 6/10–11
    shrinkable aggregates, 6/6
    steel corrosion, 6/4–5
    steel types, 6/3
    sulfate attack, 6/6–7, 11/22–3
    thaumasite attack, 6/7–8
    thermal cracking, 6/11–15
    water for, 6/2–3
Reinforced concrete structures, investigative procedures and fault finding equipment, 6/12–43
    acoustic emission testing, 6/41–2
    carbonation depth, 6/29–31
    cement content, 6/29
    chloride content, 6/27–8
    compressive strength determination, 6/31–2
    covermeter survey, 6/16–17
    high alumina cement structures, 6/31
    infra-red thermography, 6/42–3
    sulfate content, 6/30–1
    visual survey, 6/13, 6/16
    see also Half cell potential reinforcement corrosion testing; Petrographic examination; Radar profiling; Schmidt Rebound Hammer; Strength-testing of concrete; UPV (ultrasonic pulse velocity) measurement from PUNDIT
Reinforced concrete structures, repairs:
    corrosion inhibitors, 7/18–19
    expanded titanium mesh anode system, 7/8
    incipient anode effect, 7/5–6
    patch repairs of carbonation-induced corrosion, 7/3–5
        anti-carbonation paint, 7/5
        bonding and debonding agents, 7/4
        health and safety, 7/5
        phenolphthalein spray method, 7/3–4
    patch repairs of chloride-induced corrosion, 7/5
    sacrificial flame- or arc-sprayed zinc, 7/8
    see also Cathodic protection of reinforced concrete; Chloride removal (CR); Realkalization (ReA)
Reinforcement:
    durability and standards, 11/13–14
    epoxy-coated, 11/14
Reinforcement corrosion testing:
    corrosion rate measurement:
        about corrosion rate measurement, 6/47–8
        GECOR 6 device, 6/49
        linear polarization equipment and use, 6/48–50
        linear polarization interpretation, 6/50–1
    electrical resistivity measurement:
        equipment and use, 6/46
        interpretation, 6/46–7
        limitations, 6/47
    see also Half cell potential reinforcement corrosion testing
Repairs of carbonation-induced corrosion, 7/3–5
Residuals analysis, 10/31–2

Sacrificial flame- or arc-sprayed zinc protection system, 7/8
Sacrificial protection *see* Cathodic protection of reinforced concrete
Sample data and probability measures:
    critical values, 10/12
    data representation, 10/3–5
    expected values, 10/8
    histograms, 10/4
    normal distributions, 10/8–9
    population measures, 10/6
    probability/probability functions, 10/6–8
        calculations, 10/9–11
    quantitative measures, 10/5–6
    random variations, 10/2–3
    sample data considerations, 10/3
    scatter diagrams (scattergrams), 10/4–5
    standardized normal variate, 10/10–11
Sampling concrete *see* Core sampling and testing; Fresh concrete sampling; Hardened concrete and mortar sampling
Sampling and estimation:
    Central Limit Theorem (CLT), 10/13–14, 10/15
    comparison of means, 10/18–20
    comparison of variances, 10/20–1, 10/25
    confidence levels, 10/16–17
    control charts, 10/17–18
    large sample statistics (normal distribution), 10/13–14
    sample statistics, 10/13
    small-sample statistics (t-distribution), 10/14–16, 10/37
Sampling and testing management, 8/30
Sandberg cement content test *see* Pressure filter (Sandberg) cement content analysis method
Scatter diagrams (scattergrams), 10/4–5
Schmidt Rebound Hammer:
    about Schmidt hammer, 6/35
    age/hardening rate/curing type effects, 6/37–8
    basic principle, 6/35
    calibration, 6/38–9
    cement content effects, 6/36–7
    cement type effects, 6/36
    coarse aggregate effects, 6/37
    compaction effects, 6/37
    mass of specimen effects, 6/37
    moisture condition effects, 6/38
    operating procedure, 6/35–6
    stress state effects, 6/38
    surface carbonation effects, 6/38
    surface type effects, 6/37
    temperature effects, 6/38
    theory, calibration and interpretation, 6/36
Self-certification and quality control, 8/20–2
Shewhart charts:
    Action Lines (Upper and Lower Control Lines), 9/4–5

mean strength monitoring, 9/3–5, 9/8
Method of Ranges, 9/5
Runs Analysis, 9/7
standard deviation, monitoring, 9/5–6
Target Range, 9/5
trends analysis, 9/7
Warning Lines, 9/5–6
Shrinkable aggregates, in reinforced concrete, 6/6
Significant tests:
  comparison of means, 10/24
  comparison of variances, 10/25
  hypothesis testing, 10/21–3
  significance and errors, 10/25–6
Specifications:
  about specifications, 11/2–3, 11/27
  performance-based, 11/9
  prescription-based, 11/8
  production and conformity, 11/24–5
  role and status, 11/3–4
  typical clauses:
    strength sampling, 11/28
    temperature monitoring and control, 11/28
  using, 11/29
  writing recommendations, 11/29
Standardized normal variate, 10/10–11
Standards:
  about standards, 11/2
  ACI 318-02 (exposure conditions), 11/18
  ACI 318-99 (aggregates), 11/13
  ACT 318-99 (water), 11/12
  ASTM C1157-00 (cement), 11/12
  BS 1881:
    accelerated testing, 3/3–5, 3/11
    chemical analysis, 4/1–2
    compression testing, 2/4
    core sampling and testing, 5/1–2, 5/5–6
    covermeter survey, 6/17
    fresh concrete analysis, 1/4
  BS 5328 (compliance rules), 9/17–18
  BS EN 12350 (sampling), 5/3
  BS EN 12390-1 (accelerated testing), 3/3
  BS EN 12390-2 (cube manufacture and curing), 5/3
  BS EN 12390-3 (cube testing), 3/3, 5/3
  BS EN 12390-4 (compression testing machines), 2/5–6
  BS EN 12504 (cored specimen testing), 5/2, 5/5
  BSI Catalogue, 11/5
  BSI CP110 (accelerated/early-age testing), 3/13–15
  core sampling and testing, 5/1–2
  EN 197-1 (cement), 11/12
  EN 206-1, 11/7, 11/9
  EN 206-1 (additions), 11/12
  EN 934-2 (admixtures), 11/13
  ENV 1991-1, 11/10–11
  Eurocodes, 11/7
  European Committee for Standardization (CEN), 11/6–8
  ISO 14654 (epoxy-coated reinforcement), 11/14
  ISO 17025 framework, 8/29–30
  ISO (International Standards Organization), 11/6
  Jass-5 (Durability), 11/11
  limitations, 11/26–7
  performance–based, 11/9
  prescription–based, 11/8
  role and status, 11/3–5
  selection of, 11/4–5
  specific to industry standards, 8/15–16
  and specification conformity, 11/24–5
  for supporting processes, 8/14–15
  systems standards, 8/14
  worldwide use, 11/5
  see also ISO (International Standards Organization)
Statistical analysis:
  about statistical analysis, 10/1
  F-distribution critical values table, 10/38–9
  normal distribution table, 10/35–6
  selected statistical formulae, 10/33
  selected statistical techniques in ACT, 10/34
  t-distribution critical values table, 10/37
  see also Regression models; Sample data and probability measures; Sampling and estimation; Significant tests
Steel for reinforced concrete:
  corrosion problems, 6/4–5
  types of, 6/3
  see also Reinforced concrete structures, formation, problems and faults
Strain gauges, ERSG (electrical resistance strain gauge), 2/7
Strength-testing of concrete:
  about strength testing, 2/1
  BS 1881, 2/4
  comparative cube verification, 2/8
  compression testing machines specifications, 2/5–6
  compressive strength determination, 6/31–2
  Concrete Society Working Party (1971), 2/3–4
  flexural strength testing, 2/10–11
  Footemeter strain-gauged column, 2/6–7
  force calibration, 2/7
  force transfer verification, 2/6–7
  lazy tong device, 2/9
  one-side (bending) modes of failure, 2/4–5
  potential and in-situ strength, 5/3, 5/4, 5/8–10, 5/12–14, 5/16
  tensile splitting testing, 2/8–10
  uniaxial compression testing, 2/1–5
  with UPV measurement, 6/26–7
  variability problems, 2/1–5
  verification procedures, 2/6–8
  see also Accelerated strength testing of concrete; Cusum charts/technique; Reinforced concrete structures, investigative procedures and fault finding equipment
Sulfate attack on reinforced concrete, 6/6–7, 11/22–3
Sulfate content:
  determination, 4/11
  reinforced concrete:
    effects, 6/30–1
    test methods, 6/31
Sulfate-resisting cement, with reinforced concrete, 6/2

Surface hardness tests, 6/35–9
   see also Schmidt Rebound Hammer

Test methods for materials, 11/14
Test samples, fresh concrete, 1/4–5
Thaumasite attack on reinforced concrete, 6/7–8
Thermal cracking of reinforced concrete, 6/11–15
Twenty-day strength *see* Early-age and accelerated strength testing

UPV (ultrasonic pulse velocity) measurement from PUNDIT, 6/17–27
   about UPV measurements, 6/17
   accuracy, 6/20
   applications, 6/19–20
   cavity detection, 6/24
   crack depth estimation, 6/24–5
   defect detection, 6/24
   fire damage estimation, 6/25–6
   frequency of vibrations, 6/18
   homogeneity of concrete observation, 6/24
   long term change monitoring, 6/25
   shape of specimen, 6/18
   size of specimen, 6/18
   strength estimation, 6/26–7
   test conditions influence, 6/22–4
   test method, 6/18–19
   transducer coupling arrangements, 6/21–2
   transducer coupling problems, 6/21
   velocity of longitudinal pulses, 6/17–18
   void detection, 6/24

Variances, comparison of, 10/20–1, 10/25
Visual examination and measurement, 5/5–6
Voidage, 5/7, 5/13, 5/18, 5/19
   detection with UPV measurements, 6/24

Water, durability considerations, 11/12
Water content of fresh concrete, 1/19–21
   absorbed water content, 1/20
   free water content, 1/20
   high-temperature method, 1/19–20
   microwave oven method, 1/20–1
   oven-drying method, 1/21
Water quality, reinforced concrete, 6/2–3
Water/cement ratio, 4/11–12